MX

MX Prescription for Disaster

Herbert Scoville, Jr.

The MIT Press
Cambridge, Massachusetts
London, England

This book was set in Baskerville by Achorn Graphic Services, Inc. and printed and bound by Halliday Lithograph in the United States of America.

Library of Congress Cataloging in Publication Data

Scoville, Herbert.
MX: prescription for disaster.

Includes index.
1. MX (Weapons system) 2. United States—National security. I. Title.
UG1312.I2S36 358'.174'0973 81-8294
ISBN 0-262-19199-7 (hard) AACR2
0-262-69077-2 (paper)

To George B. Kistiakowsky,
always stimulating, always an inspiration

Contents

Preface

Over the past thirty years I have been exposed to almost every facet of nuclear war except, I am thankful, nuclear war itself. In the aftermath of Hiroshima and Nagasaki I studied the effects of nuclear explosions, first for the Atomic Energy Commission and later for the Defense Department. Later while working for the Central Intelligence Agency, I analyzed Soviet nuclear weapons and the military systems for delivering them and compared these with our own. Since sound technical intelligence is an essential prerequisite to nuclear arms control, I soon became involved in looking for ways by which the ever-growing dangers of nuclear war could be lessened. Then I moved to the Arms Control and Disarmament Agency, where I continued my efforts to reduce the nuclear threat not only from a confrontation between the United States and the Soviet Union but also from the spread of these weapons to additional nations or groups.

Based on my observations from all these vantage points, I am convinced that a nuclear war would be an unparalleled disaster for mankind; I am convinced that each day is carrying us closer to the outbreak of a nuclear war; I am also convinced that the only way for our civilization to avoid this disaster is to start now to bring the development, production, and deployment of nuclear weapons under control. Unless we do, one day they will be used again and the world will never be the same.

The MX missile is only one in a long line of nuclear weapons, and unfortunately it will probably not be the last. It epitomizes all that is most dangerous and bad in nuclear weapons, however, and I have chosen it as the centerpiece of this book in order to focus attention on the issues that nuclear weapons present. But we must not be misled into believing that the problems of nuclear war will be solved if we succeed in dealing with the MX. That will be only the first step, but we may have learned much in taking that step.

I first became deeply involved in the MX in November 1979 when I was invited to Salt Lake City by State Senator Frances

Farley, who inspired me, as she has many, many others, to make this issue the centerpiece of my activities for the next two years. She foresaw before most others not only the impact the MX program could have on that region of the country but, more important, the dangers it created for all people. By asking questions and by not accepting pat answers, she and her associates made the MX the subject for a broad national debate. I am truly grateful to her, her associate Chad Dobson, to Jeff Van Ee, Bill Vincent, Sylvia Baker, Joe Griggs, Glenn Miller, and to the countless others in Nevada and Utah not only for the help they have provided me in preparing this book but also for having made the MX issue come alive.

I should also like to thank my colleagues on the board of the Council for a Livable World and Jerome Grossman, its president, for having urged me to undertake the writing of this book and for providing assistance in getting the project underway. The financial aid that members of the council made to defray some of the publication costs is greatly appreciated. Without this encouragement and support it would never have come to pass. I am particularly grateful to George Kistiakowsky, Kosta Tsipis, and John Isaacs, who spent many hours reading and providing invaluable comments on the manuscript.

I have also received tremendous help with both information and comments from a host of others, and Mike Mawby and Marilyn McNab deserve special thanks. Bill Kincade and his fellow workers at the Arms Control Association continually provided important assistance in many ways. Paul Walker and Chris Paine were also extremely useful in supplying essential material.

I am particularly grateful to Ralph Ingersoll for his inspiration and advice based on a lifetime in communicating with the public.

And above all I must express my deep gratitude to Barbara Givens, who translated my verbal meanderings into written words, deciphered my unique scrawls and corrections, and then transferred them again and again and again into a readable copy. Without her tireless work, the deadlines would never have been met.

Finally, I should like to express my heartfelt appreciation to my wife, Ann, and my two sons, Tony and Tom, who have always encouraged me to keep on over the years; without them I might have faltered on the way. Even at the end they had the patience to go through the manuscript and offer many very useful suggestions to make it more intelligible and pertinent.

MX

1 The MX and Nuclear War

The true security problem . . . is not merely man against man or nation against nation. It is man against war . . . . When we get to the point, as we one day will, that both sides know that in any outbreak of general hostilities, regardless of the element of surprise, destruction will be both reciprocal and complete, possibly we will have sense enough to meet at the conference table with the understanding that the era of armaments has ended and the human race must conform its actions to this truth or die.

Letter written by President Eisenhower on April 4, 1956, Archives, Eisenhower Library, Abilene, Kansas.

No threat to the future of civilization compares with the threat of nuclear war. No national goal is more important than the goal of avoiding the outbreak of a nuclear conflict. No governmental actions require greater scrutiny than those actions that could increase the chances that nuclear weapons will be used by anyone, anywhere on this earth.

The decision to procure and deploy the MX land-based intercontinental ballistic missile is just such an action. It could have more influence on the likelihood of nuclear war than any weapon decision of the past decade. It is a weapon that can be both a means to launch an atomic strike and a magnet to attract an atomic strike against it—and against us.

The MX missile system is a weapon that can be described only by hyperbole. It is already, and will continue to be, the most controversial military program on which this generation of politicians will have to take a stand—more so than the B-1 bomber, more so than the Safeguard antiballistic missile defense system (ABM), which was the subject of a major national debate in 1969, and more so than missiles with multiple warheads (MIRVs)*, which slipped through without any major public outcry in 1970.

The MX will be the largest construction project in our history, larger than the Panama Canal or the Alaskan pipeline. It is the most expensive weapon system for which our citizens have ever

*A MIRVed missile is a missile that can carry multiple reentry vehicles with warheads, each of which can be independently aimed at separate targets.

been asked to pay taxes. Its 200 missiles could easily cost more than $100 billion to deploy. It will exact a higher price from the society and the environment in regions in which it will be deployed than any previous military project. Forty thousand square miles of nearly virgin land will never be the same again, and the construction will profoundly alter the social structure in Utah and Nevada, or wherever the MX is based in the United States.

But these costs are of secondary import beside the risks that the MX missile system can create for the future of humanity. It is the most dangerous weapon on which decisions will have to be made in the next few years. It will not ensure that the land-based intercontinental ballistic missile (ICBM) third of our strategic deterrent triad will survive a Soviet attack,* yet this is the ostensible justification for its procurement as the replacement for our existing Minuteman ICBMs. Unless it can assuredly survive, it is only a lure to invite a Soviet strike.

It is itself a first-strike weapon, which can threaten a major portion of the Soviet strategic deterrent force but which will be militarily ineffective if not used to start a strategic nuclear war. It will be an incentive for Soviet initiation of a nuclear conflict either by accident or preemption in time of crisis.

The MX is an invitation to an accelerated and endless strategic arms race with no winners, one in which we will never know whether we are ahead or behind. It will make verifiable arms limitations a fading dream.

With weapons to start and fight a large-scale nuclear war, for the first time man has in his grasp the seeds of his destruction. The MX is a prime example of a nuclear war fighting weapon that can lead us closer to the catastrophe of an atomic holocaust.

This book explains in nontechnical terms what the MX is,

*The other two-thirds of our strategic forces are submarine-launched ballistic missiles and intercontinental bombers.

what it is supposed to do, and how it might be modified by the Reagan administration. The nuclear facts of life and the strategic concepts that were the bases for the original Carter administration decision to proceed with the MX are included as a necessary preamble for a public evaluation of the rightness of that decision. To support the stark conclusions that have been laid out here, I have analyzed the MX first in terms of its effects on national security and on the risks of a nuclear war. The material and social costs can then be viewed in the perspective of our real security needs. Finally, any appraisal of the MX must include a look at alternatives: How do their costs and benefits compare with the proposed basing of the MX missile on land. All these factors help to establish the relationship between the MX and nuclear war, the "true security problem" referred to by President Eisenhower.

I The MX Missile System

2 Objectives

Why are we building the MX missile system and what are the specific military needs that it is being designed to satisfy? Basically the MX missile (MX stands for "missile experimental") is to be the follow-on land-based ICBM to our existing Minuteman missiles. One thousand of the original version, the Minuteman I, were deployed in the early 1960s, in blast-resistant silos in a crash program developed as a reaction to what later proved to be a nonexistent missile gap and to the fears that the Soviet Union could launch a disarming first strike against our strategic bombers and first-generation Atlas and Titan ICBMs. These recurrent fears proved groundless, and the United States ended up with a large lead in strategic weapons. Now twenty years later we have decided to replace Minuteman missiles with more modern, less vulnerable weapons; the MX is the candidate.

But it is often forgotten that in the late 1960s about half of the Minuteman Is were already replaced by a somewhat larger, longer-range version, the so-called Minuteman II, which carried a single warhead with an explosive force (yield) equivalent to more than 1 million tons of TNT (1 megaton or mt). Between 1970 and 1975 the remaining 550 Minuteman Is were replaced by the Minuteman III missile, which carried three MIRV warheads, each with a yield of about 170 kt (kilotons, thousands of tons of TNT equivalent). In 1978 two years after the halt of the Minuteman III deployment and with a surplus stockpile of about 50 to 100 missiles, President Carter approved the shutdown of the Minuteman missile production line.

Nevertheless, beginning in the late 1970s and continuing today, all Minuteman IIIs are being given an improved guidance system, and for 300 of these the yield of the warheads is being raised to 335 kt. This combination will give each warhead on these advanced Minuteman III missiles a high probability of destroying a blast-hardened target, such as a Soviet ICBM silo, a nuclear weapons storage site, or an underground command post. Thus the currently deployed Minuteman missiles have been continuously upgraded; they no longer resemble the original version developed in 1960. Just because we have retained the name *Minuteman* does not mean we have not deployed more modern missiles since. Nevertheless a strong outcry has arisen for the replacement of at least some of these ICBMs by a totally new missile system.

The primary military justification for replacing this still modern Minuteman system is its potential vulnerability to a Soviet ICBM attack in the 1980s. The Soviet Union has reportedly tested warheads with sufficient accuracy and yield so that each one would have a high probability of destroying a single Minuteman silo, if its performance in tests can be duplicated in an actual operation. Between 1982 and 1987 it could have enough of these accurate warheads to provide at least a theoretical threat to the entire U.S. ICBM force. "Theoretical" is stressed because the practicality of successfully carrying out such an attack would be most unlikely.

Regardless of whether such a threat is realistic, it is unsatisfactory to have a major component of a nation's strategic deterrent even thought to be vulnerable, since vulnerable forces are tempting targets and create incentives for a preemptive strike against them in times of crisis. Therefore the desire to remove the theoretically vulnerable ICBMs and leave only clearly survivable strategic forces is not unreasonable. There is, however, no urgency for such action, since the other components of the U.S. strategic deterrent cannot be neutralized now or in the near future.

The vulnerability of a missile before launch is dependent on the way it is based, and the characteristics of the missile itself are not critical to its survival. In the past survivability has been accomplished for land-based missiles by housing them in underground silos hardened with reinforced concrete to withstand the effects of a nuclear explosion at relatively close range. Improved guidance technology, however, has made further hardening a losing game and no longer practical. High-yield weapons can now be designed to be able to explode within several hundred feet of their desired aim point, too close for any silo to survive the blast.

A strategic missile system should also be built in such a way as to permit reconnaissance satellites or other technical intelligence collectors to obtain a reliable and reasonably accurate count of the number of missiles of that type actually deployed. This process is known in the jargon of international treaties as verification by "national technical means." The Soviet Union might never have much doubt on how many deployed missiles the United States had because of the openness of our military programs; however, the reverse would not be the case for American knowledge of Soviet activities. The United States cannot afford to set a precedent for deploying unverifiable systems because in the long run this will be to our disadvantage.

If any agreed limits on strategic weapons are to be achieved, it is essential that the number of deployed missiles be verifiable. Furthermore the U.S. interest in verifiability is not limited to arms control agreements, for our security depends on having a sound basis for estimating the Soviet threat under all circumstances. Again it is the basing mode, not the missile itself, that determines verifiability.

A third characteristic sought for the Minuteman replacement relates to the size of the missile and the payload it can put on a ballistic trajectory toward its target. The older ICBMs were relatively light and carried only a few warheads. Political pressures became very strong during the Strategic Arms Limitation Talks (SALT) debates for the United States to deploy large missiles with the most advanced technology available. "Bigger is better" was a simplistic criterion for the MX that the public could understand. More warheads with higher explosive yield and accuracy were also sought.

This view was stressed particularly by those who believed that the United States should have the capacity to threaten destruction of the Soviet land-based ICBM force. However, since it was contrary to U.S. policy to launch a strategic first strike, the official MX system justification referred to it as a second-strike ICBM countersilo capability. To make the procurement of these accurate, higher-yield weapons more palatable, the Carter administration also argued that we needed to have a capacity for destroying other very hard targets, such as underground command centers where the Soviet leadership might take refuge in the event of a nuclear war.

This requirement for silo-busting warheads was also reinforced by the political desire to match the Soviet Union. Secretary of Defense James Schlesinger was the first to proclaim this in 1974, a year before the Soviets had deployed their first MIRV missile and many years before they would have the accuracy needed to threaten our ICBM silos. At that time the United States was far ahead of the Soviet Union in MIRV and guidance technology. This perception of possible future inferiority became a major popular driving force behind U.S. programs designed to acquire missiles with these characteristics. It led first to the program to install new guidance and higher-yield warheads in the Minuteman III.

Countersilo capability became an important factor behind the Carter administration's decisions on the MX program. National security adviser Zbigniew Brzezinski argued that the improved strategic power of the MX would give us the ability to respond in kind and thus reduce incentives for the Soviets to initiate an attack. He also claimed that it should improve Soviet incentives to negotiate a SALT III treaty by demonstrating our determination to maintain essential equivalence, the timeworn "bargaining chip" argument that we must buy weapons to strengthen our negotiating position. Matching the Soviet Union in every weapons characteristic, even though this did not contribute to security and might even decrease it, became a political requirement, which decision makers were loath to ignore. In the long run this political factor determined the choice of design of the MX missile more than any specific military need.

3 The MX Missile

On June 8, 1979, after considerable debate the Carter administration announced the decision to proceed with the full-scale development of the largest ICBM that would have been allowed under SALT II. The Defense Department has said that if the size were any greater, the missile could not be based so that it could survive a Soviet strike. On the other hand it is still too big to be compatible for launching from the new large Trident submarines. As the problems and expenses of the MX basing continue to grow, the idea of moving all missiles to sea becomes more attractive. The lack of flexibility resulting from the large size of the MX is thus a handicap. Now the missile development program cannot be altered in such a major way without prolonged delays in the availability of an operational missile.

The MX missile chosen for full-scale development in June 1979 would have three solid fuel booster stages, be 70 feet long, and weigh 192,000 pounds. Thus, the MX will be about two and a half times as heavy as the Minuteman III that it would replace. Its diameter is 7 feet 8 inches, slightly more than 2 feet greater than the maximum diameter of the Minuteman (figure 1).

Although there was no Carter administration plan to do so, the MX could still be fitted into the Minuteman silos because these silos are somewhat larger than the Minuteman missile itself to allow space for the escape of the exhaust gases produced by the burning propellant. Such extra space would not be necessary in the case of the MX because, unlike its predecessor, it would use compressed gas to pop it out of its launcher or silo and its solid fuel first-stage rockets would not be ignited until the missile was above the ground. This is the same technique, known as cold launch, that has been used for years with our submarine missiles and that the Soviets have been using in their modern SS-17 and SS-18 ICBMs. The dimensions of the MX are approximately identical with those of their other modern ICBM, the SS-19, which was designated in the SALT II Treaty as having the maximum size allowed for any replacement ICBMs on either

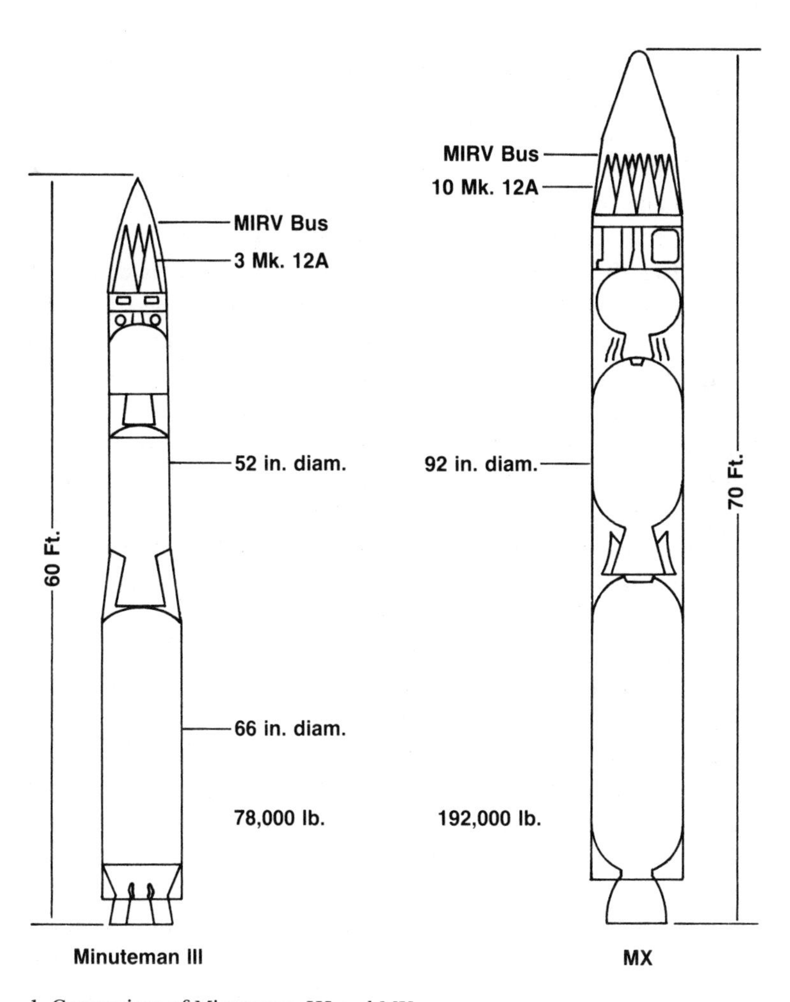

1. Comparison of Minuteman III and MX

side. It is, however, significantly smaller than the 308 very heavy SS-18s permitted to the Soviet Union. The United States accepted provisions in SALT I and II that barred our deployment of such heavy ICBMs after the Joint Chiefs of Staff found it unnecessary to keep this option open in the absence of any U.S. military requirement for such large missiles.

The fourth liquid-fueled stage of the MX missile is designed to carry ten MIRV warheads capable of being aimed at individual targets. This stage is often called the "bus" since it drops off the individual warheads (the passengers) one by one after they are on the proper trajectories to their separate targets. The explosive force or yield of each warhead in the first model MX (Mod 1) would be 335 kt; this is the same warhead that is now being installed in the modernized Minuteman III. In a later modification of the MX (Mod 2) it is planned to increase this yield still further to 500 kt, more than thirty times that of the original bomb that destroyed Hiroshima in 1945.

The guidance system for the first version of the MX, like the warhead, will be the same as that used in the more advanced Minuteman III. It is predicted to have an accuracy (CEP)* of approximately 600 feet at a range of about 6,000 miles. The later version of this missile would have incorporated an improved guidance system, which could improve the accuracy by a factor of about two (reducing the theoretical CEP to about 300 feet).

Accuracy is much more important than yield in determining the effectiveness of a warhead in destroying hard targets. If the planned characteristics are actually obtained when the MX is developed and tested, then each warhead of the initial MX (Mod 1) would have a 68 percent probability of being able to destroy ("kill") a Soviet ICBM silo even if it were hardened to withstand

*CEP stands for circular error probable, a measure of accuracy. It means that half of the warheads aimed at a given point will fall within a circle of that radius around that aim point.

a blast pressure of 3,000 psi (pounds per square inch), the maximum believed possible. Firing two warheads at the same silo would give a 90 percent kill probability, if the explosion of the first warhead did not destroy or interfere with the second. The more advanced version of the MX missile (Mod 2) would have a 99 percent kill probability with a single warhead. Fundamentally the MX missile has been designed to be a Soviet silo-buster (killer); the total 200-missile deployment with 2,000 highly accurate warheads will be a threat to the entire ICBM portion of their strategic force.

For comparison the Soviet SS-19 can carry six warheads, each with yields possibly as high as 600 kt. Their larger SS-18 can carry ten warheads with yields of about 1 mt. The SS-18 could be modified in the absence of a SALT II Treaty to carry twenty or even thirty warheads, albeit with much lower yields (perhaps 200 kt). Thus the SALT II Treaty put an important limit on the SS-18 program, which plays an important role in determining the feasibility of designing an MX system that will survive a Soviet attack. The advanced guidance system on the Soviet MIRVed SS-18 and SS-19 is estimated to provide an accuracy of about 800 feet. This gives each SS-18 warhead a probability of about 82 percent for destroying a Minuteman silo and of 98 percent for destroying an MX shelter. (Table 3.1 summarizes the probability that different U.S. and Soviet weapons will destroy certain hardened targets.)

In the aftermath of a hypothetical Soviet first strike, the Pentagon estimated we would need 1,000 warheads to destroy any Soviet missiles left in their launchers. The size of the MX system was therefore set at 200 missiles with 2,000 warheads because calculations showed that in the event of a Soviet attack with its forces available under the SALT II Treaty, only half, or 1,000 warheads, would survive, even with the proposed new invulnerable basing scheme. However, Soviet military leaders cannot ignore that before combat starts, the MX system would contain

Table 3.1
Probability of destroying missile silos

	Warhead characteristics		Target	Kill probability (%)	
Weapon type	Yield (kt)	CEP (ft)	hardness (psi)	1 warhead	2 warheads
MM III (advanced) / MX (Mod 1)	335	600	3000 (max. silo hardness)	68	90
MX (Mod 2)	500	300	3000 (max. silo hardness)	99	99
SS-18 (old)	1000	1500	2000 (Minuteman silo)	38	61
SS-18 (advanced)	1000	800	2000 (Minuteman silo)	82	97
SS-18 (future)	1000	600	2000 (Minuteman silo)	95	99
SS-18 (advanced)	1000	800	600 (MX shelter)	98	99
SS-19 (advanced)	600	800	600 (MX shelter)	93	99
SS-18 (20 warheads)	200	600	600 (MX shelter)	89	99

2,000 silo-busting warheads, which they can only view as a first-strike threat to their 1,400 ICBM launchers. This threat would be seen as even greater when it is remembered that the United States would have in addition 1,000 to 1,500 Minuteman III warheads with equivalent hard target capability and that ICBMs are the major part of the Soviet strategic deterrent force.

Full-scale development of the MX missile was begun in the spring of 1979, and the first flight test of this weapon is scheduled for early 1983 at Vandenberg Air Force Base, California. The engineering development program is on schedule, and the technology for the missile is reasonably well in hand. Furthermore, the Reagan administration has supported the development of the MX missile itself (although not the basing mode), so program modifications, which could delay the development, are not anticipated. Nevertheless past experience has shown that few weapons programs meet their predicted schedules.

Before initial deployment, twenty missile flights will be conducted to demonstrate the overall weapons system capability. The first flight tests will not be from operational-type launching systems because of the delay in making the decision as to how the missile would be based. After about five flights, testing could occur from operational-type mobile launchers. Were the decision on a launching mode to be changed again—and this is quite possible with the new Reagan administration—then complete system flight tests would be still further delayed. The entire MX missile development could, however, still be completed in time to meet the current planned initial operational date of July 1986 for the first ten missiles.

4 The Basing Mode

The Carter administration selected the basing mode for the MX in September 1979 with the twin objectives of providing for survivability and verifiability. Fundamentally the decision was to develop a mobile missile with multiple protective shelter (MPS) basing. The MX would be shuttled among twenty-three shelters, each hardened to withstand a blast pressure of about 600 pounds psi, under a shield so that Soviet satellite observation could not determine at which shelter the missile was stationed at any specific time. In simple terms the United States proposed to play the classical shell game with the Soviet Union in which the MX missile was a 190,000 pound pea and heavy concrete shelters were the shells. On top of the shelters were to be ports that could be opened on request so that Soviet satellite cameras could verify that there was only one missile in each cluster of twenty-three shelters.

At intervals of several months, or when maintenance was required, the missile and its launcher would be moved in a horizontal position on a transporter and under a shield along heavy-duty roads connecting the various shelters. The total hardware package of missile, launcher, transporter, and shield would weigh about 1.6 million pounds. When the plan was first announced in 1979, it was proposed that both the missile and its launcher would remain on the transporter under the designated protective shelter. During routine movements only the shield would travel back and forth from shelter to shelter to confuse the Soviets as to where it had deposited the missile. Upon warning of a Soviet attack, the missile could then dash at 30 mph, unmanned and by remote control, to a new shelter before the warhead from the Soviet Union could reach the launch complex. This was considered the ultimate safety factor in the event that the "shell game" deception broke down and the Soviets somehow learned which shelter held the missile.

Six months after the initial decision, the Defense Department revised the basing plan in order to cut costs. Only the launcher

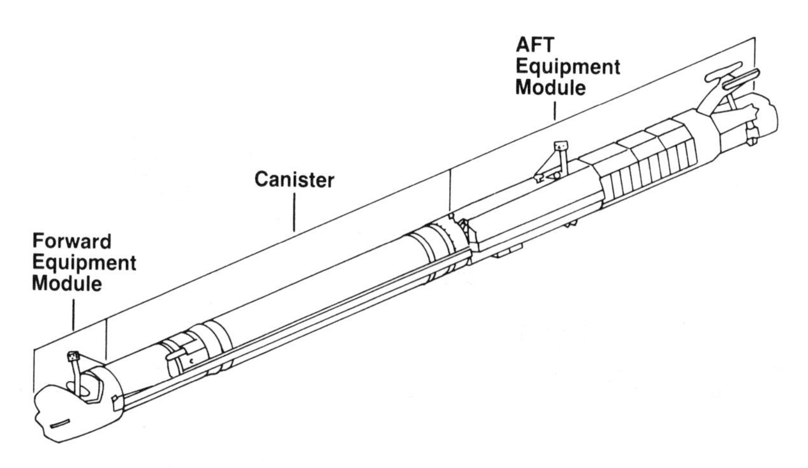

2. MX launcher: 155 ft long, 110 in. in diameter, 500,000 lb.

and missile would be placed under the shelter, and the transporter with the shield as an integral part would remain in the open. Since the transporter was larger than the rest of the system (figures 2, 3), the shelters could be reduced in size, making it easier to harden them to withstand nuclear explosions and thus saving money. This was not the only change. The original plan called for the missile to be elevated, break through the roof of the shelter, and then be fired without being taken out of the shelter. This procedure was altered to require the missile and launcher to move partially out from under the shelter roof before elevating the missile and firing it (figure 4). This also made the shelters easier to build. The launcher and missile would no longer be able to move from shelter to shelter by themselves, however, but must be loaded on to the transporter before acquiring that mobility.

When the actual missile with its launcher is left behind in a shelter, the new plan called for replacing it on the transporter and under the shield with a dummy, which would simulate the observable characteristics of the missile and its launcher. This

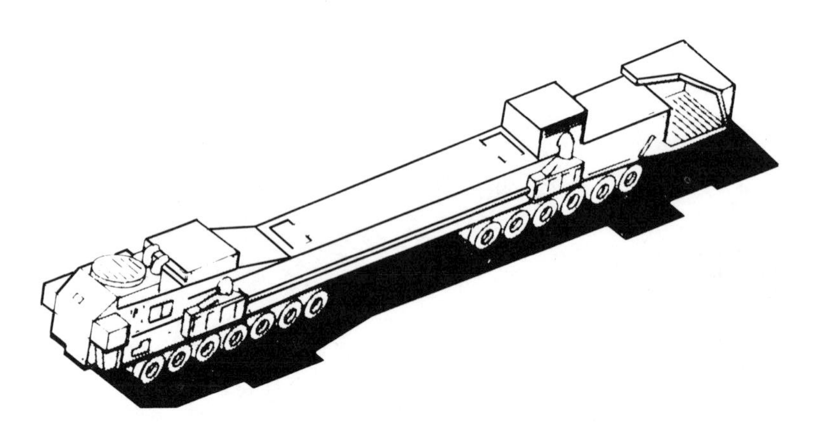

3. MX transporter (used inside cluster): 201 ft long; 25 ft wide overall, 16 ft wide over tires; 31½ ft high; 1,600,000 lb loaded. Shield to conceal missile is integral with transporter.

operation by itself does not promise to be easy to carry out without leaving evidence. The dummy must have the same weight as the missile and its launcher, and it must have simulators to reproduce the other more subtle characteristics of the missile that ouside observers might use to determine whether it was on the transporter. For example, the missile in its canister must be kept air-conditioned at a constant temperature in order to be prepared for firing on a short notice. This temperature differential from an empty shield would be detectable by sensitive infrared sensors. Thus the dummy must reproduce these infrared characteristics. It must also duplicate any acoustic or seismic signals or nuclear radiation that might be given off by the missile and might be detectable outside the shield. Since the gross weight of the missile with its launch vehicle is one-half million pounds, the dummy with its simulators will be a significant part of the entire MX hardware package. Furthermore, there will have to be twenty-two dummies in each cluster so that one will be available for substitution in any shelter in which the missile is left behind.

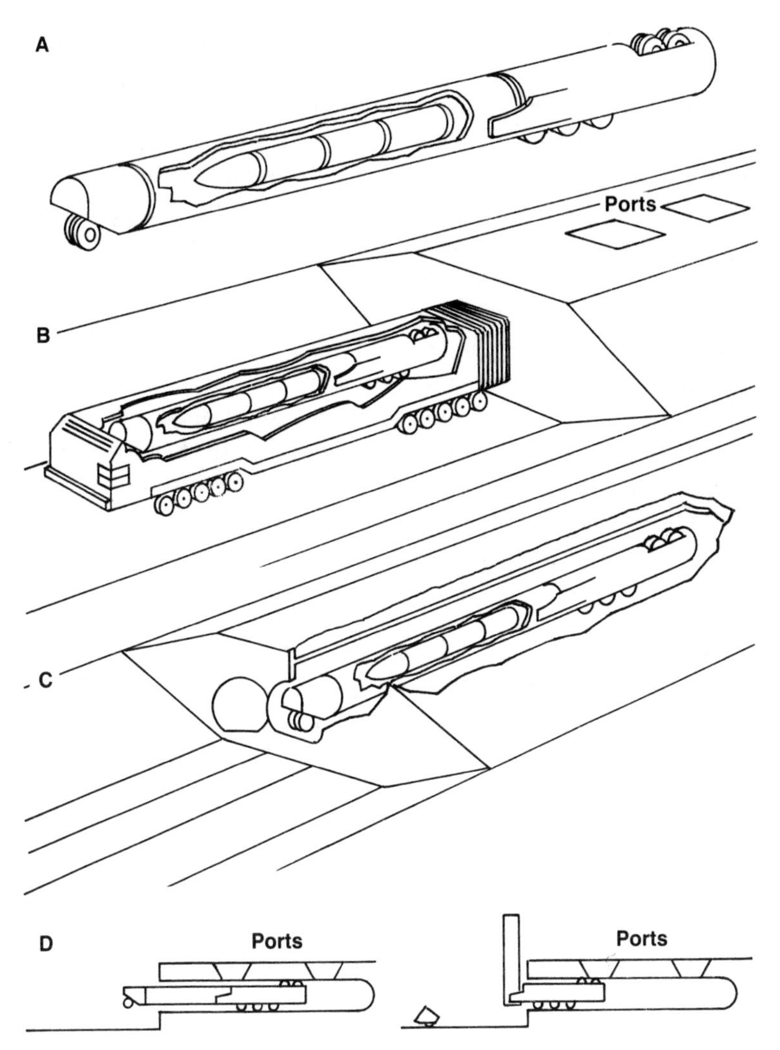

4. MX launch sequence (horizontal MPS system). A. Missile on launcher. B. Transporter with missile and launcher outside shelter. C. Missile and launcher in shelter. D. Launcher and missile outside shelter, ready for launch. Shelter is 180 ft long and 15 ft in diameter. Ports would be opened on request to permit Soviet satellites to verify whether shelter contains a missile.

It is not always easy to practice deception in this age of advanced intelligence technology.

The change in design that separated the transporter from the missile and its launcher had one serious drawback; it compromised the important feature of allowing all missiles to dash from one shelter to another in the event that the shell game deception failed. Now the transporter would have to be at some distance from the shelter that houses the missile to avoid disclosing its location. To change shelters, the transporter would now have to move back to the missile shelter, off-load the dummy, load the missile and launcher, and carry them several miles to another shelter. This operation could never be done in the available fifteen minutes warning of an attack; instead several hours will now be required to relocate the missile. In order not to forgo completely a dash capability, the new plan calls for a small percentage, perhaps 10 percent, of all the missiles to be kept on the transporter out in the open, perhaps near or in the maintenance facilities. These could then dash to any one of the twenty-three shelters in its cluster within the fifteen-minute warning time. The disadvantage is that missiles that are on the transporter in the open are not protected from a nuclear attack and are vulnerable to short-range submarine-launched missiles in much the same fashion as are bombers on the ground. Since deception of the Soviet Union is an uncertain long-term proposition, it is unfortunate that in the first six months of its existence, some of the dash characteristics, the MX system's extra safety factor, had to be sacrificed to escalating costs.

As originally designed, each missile would be in a cluster with twenty-three shelters connected by roads, which could carry the 1.6 million pound missile-launcher-transporter system. The entire 200 MX system would require 4,600 shelters and about 8,500 miles of roads, more than the distance from Maine to California and back. This number, 4,600, was selected to ensure the survival of half of the missiles even if the system were attacked by virtually the entire Soviet ICBM force allowed under the SALT II Treaty.

In September 1979, the original Carter plan placed the twenty-three shelters on short side roads off a large loop road, nicknamed "racetrack" (figure 5A). The minimum separation between shelters was chosen to be approximately 7,000 feet in order to ensure that one Soviet warhead could not destroy more than a single shelter.* For 7,000 feet separation between shelters, the area covered by each racetrack cluster would be greater than 30 square miles.

In the spring of 1980 this racetrack design was modified to a series of linear clusters with straight roads off which the shelters would be placed (figure 5B). Linear roads might facilitate somewhat the location of the sites within actual terrain, but they should not significantly reduce the total area required in the relatively flat valleys of Nevada and Utah, the location favored by the air force. The area covered is determined by the number of shelters and the maximum separation chosen to ensure that only one can be destroyed by a single warhead. Only by packing the shelters closer together can the area be significantly reduced. The roads are only the connecting links between shelters, and whether they are in loops or straight lines should not materially alter the total area covered by a single missile cluster despite air force claims to the contrary. In order to reduce land use, the minimum separation between shelters was reduced from 7,000 feet to about a mile, thus reducing the safety factor.

The new configuration does make it somewhat easier to expand the system if subsequently it is determined that the 4,600 shelters are not sufficient to ensure survivability; indeed this is already proving to be the case. With the closed-loop design it

*The diameter of a circle within which a 1 mt warhead can destroy a 600 psi target is approximately 3,600 feet. However, electromagnetic radiation from such an explosion, which could damage electronic components of the missile system, makes a separation of less than 5,000 feet unsafe.

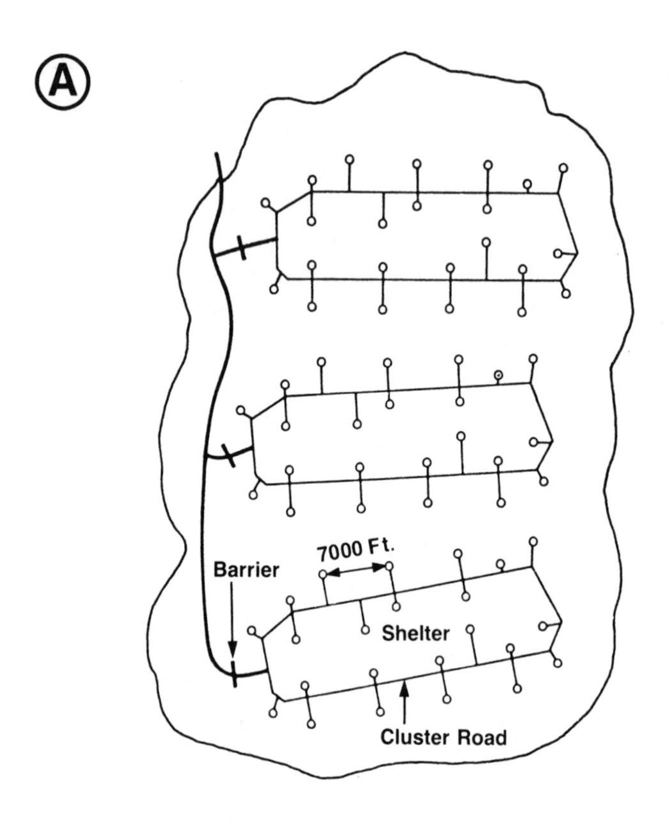

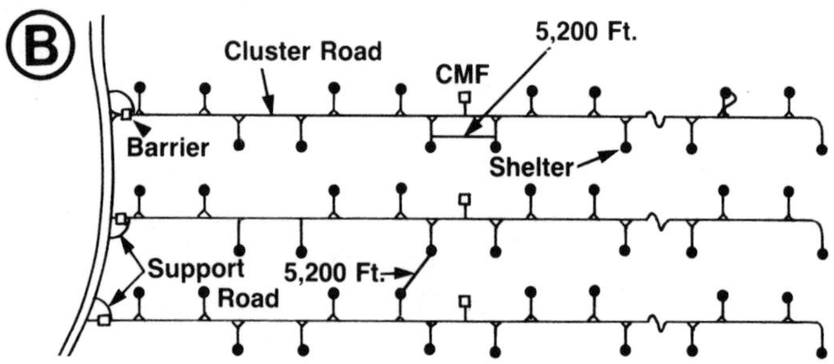

5. A. Loop shelter cluster ("racetrack"). B. Linear shelter cluster, with 23 shelters per cluster. CMF: cluster maintenance facility.

could be difficult to add additional shelters, but with an open-ended linear one, expansion would be much easier. In the long run this change, which was advertised as making the system more acceptable to the residents of Nevada and Utah, may just be a way of making it easier eventually to cover larger acreage with more missiles and shelters.

An important characteristic of every missile system is its command and control and the communication with the missile launch mechanism. (These elements are often known as C^3.) A missile must not be launchable without a specific signal from higher command. No individual at a missile launch complex should be able to start a nuclear war on his own. The systems must be fail-safe; if any failure in the launch controls occurs, the mechanism should prevent the launching of missile rather than allow it to be fired. Elaborate procedures, which must remain secret, have been adopted in the Minuteman system in order to avoid this possibility. On the other hand, in order to be able to have such tight, high-level control and still retain a credible deterrent, the command, control, and communications mechanisms must also be able to survive a nuclear attack. Otherwise the MX system could be put out of action by a Soviet attack on its command structure or its communications instead of on the missiles themselves.

Redundancy is an important characteristic of a survivable control system. Land lines, multiband radio, air, and space systems—all must be utilized. Since land-based command posts, even those deep underground, can be put out of action by highly accurate missiles, these by themselves are not sufficient. Placement of alternate command centers in aircraft, some of which are constantly in the air, is a more satisfactory way of ensuring survivability since these cannot be located at any specific moment. The enemy would never know where to aim in order to try to destroy the aircraft. However, radiation emitted by a nuclear explosion in the cluster area could temporarily interfere with radio communications.

The missile launcher cannot be manned because this would give away the location of the shelter with the missile. Manning all twenty-three shelters of each missile cluster twenty-four hours a day would be prohibitively expensive. Therefore the command message to fire must be fed directly into the MX launch mechanism from the command aircraft. Furthermore, since currently there is a desire to have the capability for a flexible response—that is, to retaliate in alternative ways dictated by the nature of the aggression—last-minute retargeting of the warheads must be possible. The entire command and communication structure can become very complicated. It is not enough to send a simple order to fire. A more complete message must be sent, and communication in most cases must be two-way so that the control knows the operation is going forward satisfactorily. Instructions to the guidance computer to alter the warhead's target may have to be sent to the missile after it has been launched. The details of the command and control techniques and operations that will actually be employed for the MX are classified, but even an outside observer can realize that the achievement of survivable, reliable, and flexible command and control is one of the most difficult technical problems the MX missile system designers have to face.

The multiple protective shelter scheme was designed at great expense primarily to fool the Soviets as to which shelter held the MX. At the same time it was important for verifiable arms limitations to provide demonstrable proof to Soviet satellite cameras that there was only one missile in each shelter cluster. These two seemingly incompatible requirements had to be resolved. The following features were included in the system design in order to support verification without compromising its basic deception characteristics:

1. Removable viewing ports would be provided on the shelter roofs. Upon request these ports would be opened so that satellite cameras could look into the shelters and determine that there was only one missile per cluster. The number of shelters opened for inspection at any one time would be small enough not to compromise the location of a significant fraction of the MX missiles but large enough to have a good probability of detecting extra missiles.

2. Only one missile would operate in a single cluster with twenty-three shelters; the only road to the outside would be blocked by barriers after the missile is in the cluster. Thus missiles from adjacent roads could not be readily moved in from the outside as a means of circumventing monitoring.

3. The maintenance, fabrication, and assembly area was designed to conduct as much construction and repair work as possible out in the open to facilitate monitoring by satellite photography.

4. The shipment of assembled equipment from the assembly areas to the deployment areas would use special transporters moving on identifiable roads at a slow pace. This procedure was designed to make it easier for satellites to spot any forbidden activity.

The first two of these four items add several hundred million dollars to the program costs; the second two cost much less. The total funds for these verification measures would be less than $1 billion (3 percent of the system cost), not unreasonable for establishing a precedent for verification of arms limitation agreements, which could enhance security and save money in the future.

The Soviet Union has never publicly stated whether it would consider such a system to be in accord with the SALT II Treaty verification provisions that forbid concealment. It would seem that the U.S. government could make a plausible argument that with these procedures it would not be practicable for us to hide a significant number of additional missiles. Opening the ports on top of the shelters when a Soviet satellite was overhead should provide a satisfactorily high probability of detection of any extra forbidden missile, although the presence of the dummies could create a problem with false alarms.

It must be kept in mind that the U.S. goal is to verify Soviet compliance, however, and it is not clear whether a Soviet multiple protective shelter scheme could be made verifiable. It is not enough for us to be sure that our system is verifiable; we must not set a precedent that can be used by the Soviets to deploy unverifiable systems.

The air force has carried out extensive studies of the best geographical location for the deployment of the MX system. When earlier designs were considered, it appeared that the western plains might be most suitable. However strong opposition from officials in Kansas and Nebraska combined with the requirement to use large areas of prime farmland caused these locations to be temporarily discarded. By the time the decision was made in September 1979 to deploy the two hundred racetracks, the air force had settled on eastern Nevada and western Utah as the most suitable location (figure 6). Later under pressure from local politicians and citizenry, the air force agreed to consider additional areas in west Texas and near Clovis, New Mexico. Some of those advising President Reagan have urged the conversion of Minuteman sites in North and South Dakota, Montana, and Wyoming-Nebraska-Colorado to the MX (figure 7).

Each missile cluster will cover on the average about thirty square miles so that a total of at least 6,000 square miles will be required just for the two hundred missile clusters. If, as the air force proposes, they are all placed in Nevada and Utah, the system as a whole will be spread over about 40,000 square miles. The MX system would cover most of the central part of these

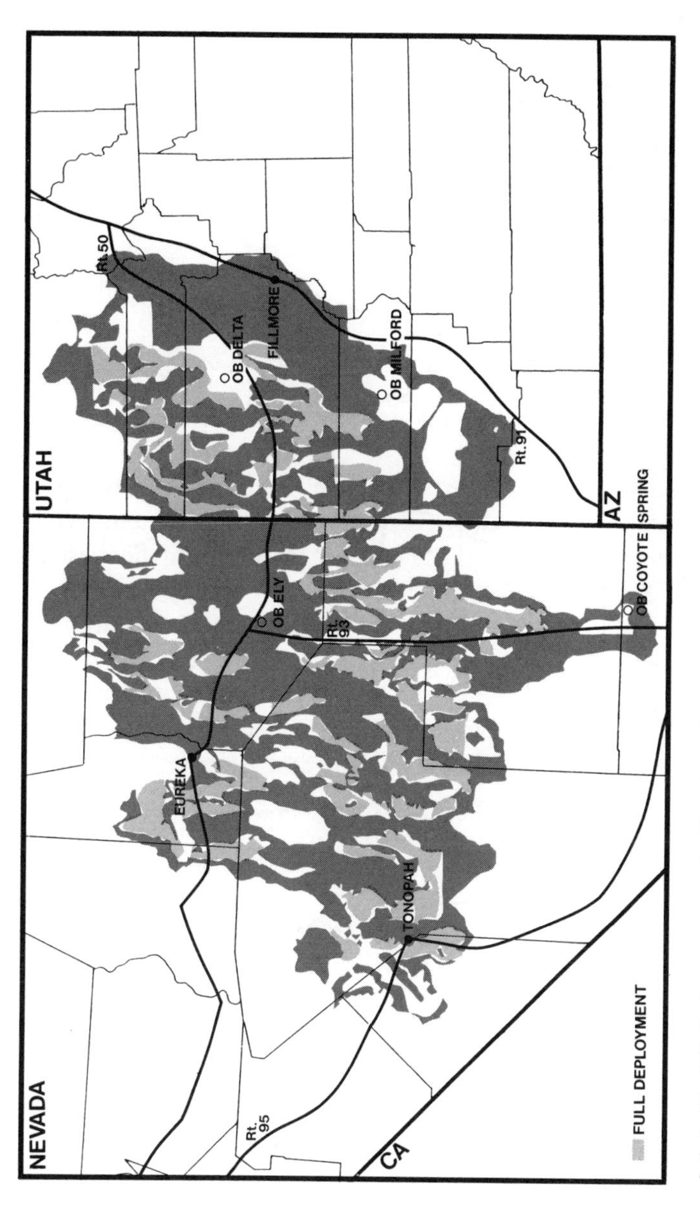

6. Suitable MX deployment areas in Nevada and Utah.

7. U.S. strategic bases: ● operational SAC bomber bases; △ ICBM fields. Crosshatched areas are candidate MX areas.

states. Not surprisingly the proposed deployment has aroused great concern in this region. If parts of the system were moved to New Mexico and west Texas or elsewhere, the impact on Nevada and Utah would be lessened, but it is not easy to find low-population-density areas of sufficient size and proper geological structure to deploy even a fraction of the MX system. Many families would have to be relocated if it were moved to New Mexico and west Texas.

Whether the September 1979–May 1980 designs for the MX deployment will remain unchanged is uncertain. As the complications and the costs of a horizontally mobile, verifiable missile system have grown, alternative approaches are being reconsidered. Specifically a concept of multiple vertical launch points or silos between which the missiles can be moved slowly in an upright position is attracting renewed attention. Some of President Reagan's advisers have urged that this type of system be built in the existing Minuteman fields. All of these alternatives will create new problems of their own and may not solve the ones in the Carter administration's plan.

II The MX Decisions of 1979

5 Realities of Nuclear War

The 1979 decisions to procure the MX missile and to build this gigantic installation in the isolated valleys of Nevada and Utah were made only after protracted consideration of a wide range of alternatives and their consequences. The decisions were evolutionary and made after years of study in which the complex mixture of technical, military, and political factors was weighed. Yet despite an avalanche of analyses and years of consideration, the final decisions were made primarily on political rather than rational military grounds, and they satisfied almost no one. Nevertheless it is important to be familiar with this background if the decisions are to be understood and evaluated.

The Catastrophe of Nuclear War

The fundamental backdrop to an understanding of our security policies in the nuclear age and to an evaluation of the decision on the MX missile system is a realization of the catastrophic consequences of a conflict in which nuclear weapons are used. Without an appreciation of this, we can be trapped into accepting the simplistic pre–World War II solution to international problems that more and better weapons mean more security. Conventional (nonnuclear) weapons are meant to fight with, and the more a nation had, the better its chances of victory. Since actual fighting with nuclear weapons will be a catastrophe for all concerned, more such weapons do not equate with greater power and can instead increase the risk of their use, with all the consequences that would entail. This is a fundamental consideration that must be kept constantly in mind in evaluating the MX program.

The world security structure was altered with the first nuclear explosion on July 16, 1945, in the Alamagordo Desert, which was called by the sixteenth-century Spanish explorers *Via del Muerte* ("the road of death"). With that initial step man had moved into a position of being able to destroy by a single action the civilization that he had spent thousands of years building. George Kistiakowsky, a builder of that first weapon, said after witnessing the explosion, "I am sure that at the end of the world—in the last millisecond of the earth's existence—the last human will see what we saw." This reality has not been widely recognized, has too often been subconsciously relegated to the unthinkable.

The first and so far only military uses of nuclear weapons occurred at Hiroshima, Japan, on August 6, 1945, and three days later at Nagasaki. The explosive force or yield of the Hiroshima bomb, which employed enriched uranium-235 and whose design had never been tested prior to its actual use, was equivalent to the energy released by the explosion of 14,000 tons of TNT (14 kt). The Nagasaki bomb, which employed plutonium and used a different and more innovative implosion principle, was identical to that tested on July 16 in New Mexico. It had a yield of 20 kt. The casualties and damage in the two cities were overwhelming; a few days later World War II was over. But the surviving Japanese living in these two cities continue to suffer the consequences. The most dramatic effects were the terrible burns caused by both the direct heat from the explosion itself and the widespread fires that rampaged through the rubble of the buildings destroyed by the blast. More than 100,000 people died in Hiroshima alone. At Nagasaki the number of casualties was lower because of the terrain on which the city was built, but the consequences were still cataclysmic.

One additional effect was introduced for the first time into warfare by these two explosions: nuclear radiation. Cities had been laid waste before by blast and fires set by conventional munitions—Dresden and Tokyo are two—but until Hiroshima no weapon had killed with nuclear radiation. About 20 percent of the deaths in Hiroshima were due to the prompt gamma and neutron radiation released at the instant the bomb went off. The number of radiation victims might have been higher had not many died earlier from burns and blast injury. Nuclear radiation kills slowly. Its victims died several weeks to a month after the explosion, wasting away with blood deficiencies and infection, which their bodies were no longer able to combat. Many died years later from leukemia.

Even so the inhabitants of these two cities were spared some of the worst consequences of a nuclear conflict. The weapons dropped on both Hiroshima and Nagasaki were exploded about 1,800 feet up in the air so that their fireballs never reached the surface of the earth. Under such conditions the radioactive fission products, residues from the explosion, never came directly in contact with large, heavy dirt particles from the ground. Therefore when the radioactive debris was carried up in the cloud to 40,000 to 50,000 feet, it did not settle back down to the earth's surface until it had drifted far from Japan. Thus the people of Hiroshima and Nagasaki were exposed only to the radiation released at the instant of explosion and were not forced to live on in a radioactively contaminated environment.*

The horrors of Hiroshima and Nagasaki did not bring to an end the development and procurement of more nuclear weapons. While there were many well-meaning attempts, and some less sincere ones, to establish controls on nuclear weapons through United Nations channels, such as the Baruch Plan, these never came near to fruition. When the Russians detonated their first nuclear explosive in 1949, the race for more, bigger, and better weapons was on in earnest; the human consequences

*There may have been some local areas with low levels of contamination, either from the reactions of the bomb's neutrons with chemicals in the soil directly beneath the explosion or from a small fraction of the radioactive debris becoming associated with some larger particles that fell to the earth nearby. However, less than 1 percent of the bomb residues came down locally, and the highest radiation levels measured were less than that allowed under normal peacetime nuclear activities. Recovery operations in the two cities were not hampered by worries over radiation exposure. Nevertheless now there is still controversy over whether some cancers that appeared twenty years later in some occupation troops cleaning up the area may have been due to this cause.

of the Japanese events were almost forgotten. Both countries began extensive programs to develop new, more efficient, and eventually much higher yield weapons. The effects of nuclear explosions were extensively studied by teams of scientists. These in turn generated a plethora of technical reports, but even though much of this information was not classified, the realities of a nuclear war faded from the public consciousness.

Radioactive Fallout

On March 1, 1954, a new dimension in the dangers of nuclear war was unveiled. At 6:45 A.M. a 15 megaton (mt) hydrogen bomb (1,000 times the explosive force of the Hiroshima bomb) was exploded on a reef of Namu Island of Bikini Atoll. The evening before at the weather briefing held to decide whether to proceed with the test, it had been predicted that any radioactive fallout would be carried by the winds off to the northwest over the open ocean and away from any land or people. But during the night the winds had begun to veer into patterns that would bring the fallout more and more to the east. In those days concern over radioactive fallout was far less than it became later and is today. The one previous thermonuclear explosion, which had been conducted two years earlier in the same general area, had given no evidence of heavy local fallout because the cloud had moved as predicted out over the ocean where it was impossible to make any measurements. Overconfidence created by this event led officials to be more casual when the winds for the March 1 explosion threatened to move the fallout pattern over inhabited islands. No decision was made during the night before the test to postpone it.

The task force ships, with all but a few persons in the firing party, were at sea about ten to twenty miles southeast of the Bikini Atoll and thirty miles from the explosion when it was set off (figure 8). Everything appeared normal until twenty-two min-

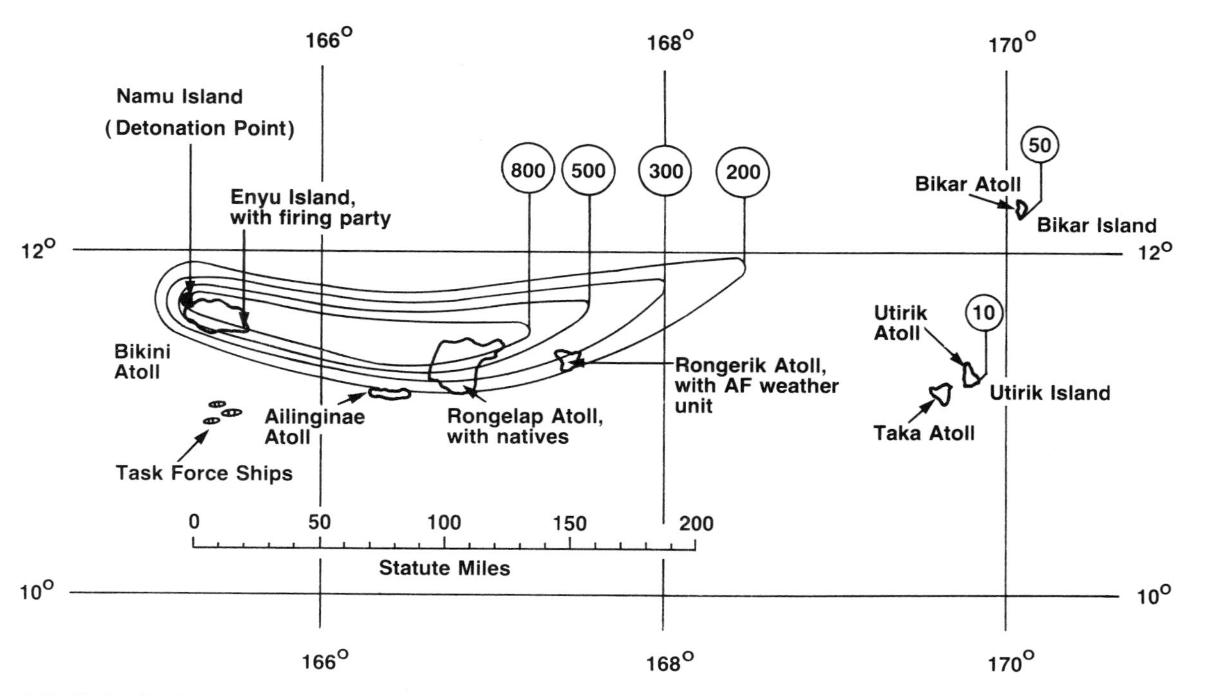

8. Radiation isodose contours for Bikini Baker test (15 megatons), March 1, 1954. Circled numbers indicate doses (in rems) that would have been received over approximately 48 hours without shielding. Survival is probable below 200 rem, unlikely above 800 rem. USAEC Report.

utes later when the small firing party on Enyu Island at the southeast corner of Bikini Atoll, and twenty miles from the detonation point, reported rapidly rising radiation readings outside the control bunker. All of the party were brought inside and the doors tightly sealed, but the radiation levels continued to rise even within the shielded building. By 8:00 A.M. radioactive material had started to fall on the ships of the task force, which were then ordered to speed to a point fifty miles south of the atoll. The washdown systems were activated and all personnel kept off the decks. People on some ships received the then maximum permissible exposure for test personnel of 3.9 roentgens, but this was not sufficiently high as to cause serious harm or concern. By 11:00 A.M. the fallout cloud passed east of the task force, and the ships moved back within ten miles of Bikini. The fallout was looked on as a nuisance but not as a cause for great alarm.

That evening radio reports received from an air force weather unit stationed at Rongerik Atoll about 180 miles to the east said that their dosimeters designed to measure the personnel radiation exposure had gone off scale at 5 roentgens. The initial reaction at the task force headquarters was one of skepticism, and the air force commander radioed back that they should not be alarmed; he speculated that their instruments were probably defective. The next day, however, realization began to set in that the fallout was far more serious than had been thought. On the day after the explosion, a full-scale evacuation of the downwind region was undertaken. Destroyer escorts were sent to Rongelap, about 120 miles to the east and most directly in the fallout path, in order to remove the Marshallese natives there. The weather unit and the natives on other nearby atolls were also evacuated. These operations were carried out by the ships most readily available, and no systematic measurements were made of the contamination levels. The natives had no radiation-measuring instruments so the extent of their exposure was not then known.

A thorough radiation survey of the contaminated areas undertaken five days after the explosion made it possible to calculate the exposures that the Marshallese and the air force weather unit had received. Measurements on the island on which eighty-two Rongelap natives lived showed that they actually received between 100 and 200 roentgens, an exposure that could make them sick but not kill them. They suffered burns from the radioactive dust, nausea, and other nonlethal radiation effects, but there were no deaths directly attributable to the radiation. Some ten to twenty years later about a third of the population, and nearly all children less than five years old, developed thyroid abnormalities. Some of these proved cancerous. Radioactive iodine, one of the fission products, had been selectively absorbed in the thyroid glands.

The survey team found that had the inhabitants been one mile farther north, where they often fished but did not live, they would have received about 400 to 500 roentgens, and half probably would have died. Had they been on the northern rim of the atoll, also about 120 miles from the explosion, they would have received 5,000 to 10,000 roentgens; all would have been killed. A one-degree additional shift in the wind during the night before the test would have meant many deaths. When the survey was all over, the task force was able to breathe a sigh of relief, but it was also able to realize how close to a major disaster the test had been.

But the task force was not yet off scot-free. Unknown to anyone at the time a Japanese fishing vessel, the *Lucky Dragon,* was sailing to the east of Bikini Atoll. Exactly where it was at the time of the explosion is still not precisely known, but it was probably about eighty miles away. When the ship returned to Japan, the crew reported having been blanketed with coral ash that was found to be heavily contaminated. One of the crew members subsequently died. The rest survived because much of the radioactive debris on the ship was washed over the sides, and the radiation levels below decks were less than those in the open.

In March 1954 the United States and the world had been given a graphic demonstration of the dangers of radioactive contamination from nuclear weapons. More than 7,000 square miles were covered with lethal fallout from this single explosion. Much larger areas were covered with lower radioactivity levels but nevertheless far in excess of those permitted even near nuclear plants in the United States. The world was provided a lesson without a major calamity. It was lucky that it got by with only a single death.

Did the world take this lesson to heart? No. A year later, in November 1955, the Soviet Union tested its first true thermonuclear weapon (hydrogen bomb) over the Arctic island of Novaya Zemlya. This was an air-burst explosion minimizing local fallout, and the radioactive cloud passed east over northern Siberia and onward all around the world. There have been no published reports of any serious ground contamination from this explosion. In the ensuing years both the United States and the Soviet Union continued to test large and small hydrogen and fission weapons, the largest being a Soviet bomb with the explosive force of 60 mt (the Soviets purposely reduced the yield from 100 mt to ensure the safety of the plane that dropped the bomb). Both countries also continued to stockpile increasing quantities of H-bombs, so that by the mid-1960s each side had more than enough weapons to produce catastrophic damage to the other under any circumstances. But they still did not stop procuring more weapons; by the end of 1979 United States had about 10,000 and the Soviet Union about 6,000 earmarked for strategic purposes alone.

In the past year we have had another event that should remind us of the dangers of a nuclear war and bring them closer to home. On May 18, 1980, the Mount St. Helens volcano erupted with an estimated energy release of the order of 10 mt, about the same magnitude as that of the 1954 Bikini hydrogen bomb. Fortunately Mount St. Helens was in a relatively unpopulated area so the casualties were low, but the entire nation was astounded by the pictures of the devastation wrought in the nearby forests. Trees were felled like matchsticks over an area of 150 square miles and in a 90 degree sector out to fifteen miles from the volcano.

Perhaps even more impressive was volcanic ash spread over tens of thousands of square miles of Washington, Oregon, Idaho, and even farther away in Montana. A week after the eruption it rained in Boston, and a film of ash was visible on cars the next morning. Although the total quantity of ash from the volcano was greater than that from an equivalent nuclear explosion, the pattern of its distribution could be the same. Fortunately the Mount St. Helens ash was mostly a nuisance and relatively benign. Imagine if it had been contaminated with trillions of curies of deadly radioactive materials, as was the coral ash from Bikini that settled on the *Lucky Dragon* and the downwind islands.* Millions would have been killed, and the entire Northwest would have been hazardous to live in for hundreds of years.

No Care, No Recovery

If a nuclear conflict broke out in which a significant portion of available weapons were used, the consequences would be incalculable. Decades and perhaps a century would have to pass before the world would recover. Humanity might not be completely destroyed, but civilization as we know it today, and par-

*One curie is radioactively equivalent to that emitted by one gram of radium (3.7×10^{10} disintegrations per second). One millionth of a curie (1 microcurie) of the fission product strontium-90 in the human body could cause an increase in the likelihood of cancer.

ticularly society in the northern hemisphere, which contains a large fraction of the developed countries, would be unrecognizable. In the United States alone the fatalities in the first few days could exceed 150 million—a number that can have no reality in our minds. The survivors—injured, burned, suffering from radiation exposure and with no medical facilities, few doctors, and only scattered supplies for their care—would envy the dead.

A few years ago a fire in a New York suburb severely burned about thirty people, yet medical facilities in the entire New York area were inadequate to care for them. Brooke Army Hospital in Texas is the only facility for treating many severe burn patients; burn casualties such as those in the 1980 aborted Iranian hostage-rescue mission are often airlifted there because of limited facilities elsewhere. But in the aftermath of a nuclear conflict, there would be no airlift; few medical personnel or hospitals would survive to provide even rudimentary care. The rare surviving doctor could only give limited attention to a tiny fraction of the injured.

Remember the problems of rescue and recovery in Italy after the 1980 earthquake despite the availability of outside aid from the entire world. Wounded burn and blast victims from a nuclear holocaust would lie and rot where they were, indistinguishable from the dead. The hundreds of thousands of dead bodies in a metropolitan area in themselves would provide a continuing hazard to the health of the survivors. Burial would be a dangerous operation.

All of this devastation would be compounded by widespread radioactive contamination. More than 7,000 square miles were contaminated after the single 15 mt explosion in 1954. In a nuclear war thousands of megatons would contaminate vast areas of the United States, the Soviet Union, and Europe. Lower-level radioactivity would spread over the entire world. Those who managed to obtain shelter before the fallout arrived might survive the early effects of the radiation, but they would eventually be forced to move about in hazardous areas. Vast populations would receive exposures far in excess of those not tolerated today even by individuals working in an atomic power plant. It would be impossible to avoid eating contaminated food or inhaling radioactive dust, which would later lodge in the bones and other parts of the body. The incidence of leukemia would rise many times. In future years large numbers of people would develop thyroid tumors as did the natives from Rongelap. In wartime they could not be evacuated outside the fallout path; they would have to continue scratching for survival in a radioactive wasteland, succumbing to the delayed effects of nuclear radiation. At a February 1980 conference in Boston, Dr. Howard Hiatt, dean of public health at the Harvard Medical School, concluded that the use of nuclear weapons would produce a medical situation resembling an epidemic for which there is no cure; his only solution was prevention. Many other physicians supported and expanded on this view. The realization of the dimensions of this disaster has since spread throughout the medical community in this country and around the world.

The total structure of our society would be altered. The damage would be so widespread that there would be no outside support to turn to in order to promote recovery. Transportation as it currently exists would be virtually nonexistent, communications would be destroyed—no long distance telephone calls to your parents or your children to tell them that you were alive, no central power to heat and light your homes. You would not have to worry about keeping your thermostat below 68°. Food would be hard to come by, and it would be impossible to ensure that it was free of radioactive contamination. You would be protecting your scarce food supplies rather than the silver with which to eat it.

But even after some semblance of social order had been restored and some steps had been taken to recover, the long-term effects would still be taking their toll. The number of neighbors and friends dying of cancer would increase radically. Sterility would take care of the population explosion. Genetic damage would be widespread not only in the next generation but in many generations to come. Finally the injection of radioactive clouds into the upper atmosphere might significantly deplete the ozone concentrations there, allowing dangerous ultraviolet rays of the sun to penetrate to the earth's surface. An increase in incidence of skin cancer could extend throughout the world. It is not possible to predict accurately the true consequences of these and other long-term environmental effects, but there is no doubt that in the aftermath of a nuclear war the world would be radically changed and few people would be spared its horrors.

Limited Nuclear War

Although the American public has been bombarded with propaganda that the Russians are planning a surprise nuclear attack on U.S. missile sites, a nuclear conflict would probably not start with a planned major nuclear strike by either the Soviet Union or the United States. Although the risk that such an attack would occur may be growing as the MX and similar Soviet weapons are deployed, the catastrophic dangers of a nuclear war are probably well enough recognized by leaders of both countries that they would not feel that any achievable superiority in weapons would warrant initiating a full-scale nuclear aggression. On the other hand a feeling of insecurity or vulnerability could increase pressures to launch a first strike with nuclear weapons.

A nuclear conflict would be more likely to break out with some very limited use of nuclear weapons, perhaps even by a third country. Unfortunately once the barrier between the use of conventional and nuclear weapons is jumped, no one knows how to keep the conflict limited. President Carter on July 12, 1977, warned that "the first use of atomic weapons might lead to a rapid and uncontrolled escalation in the use of even more powerful weapons with possibly a worldwide holocaust resulting." Secretary of Defense Harold Brown in his 1980 Annual Report said:

> We have no more illusions than our predecessors that a nuclear war could be closely and surgically controlled. There are, of course, great uncertainties about what would happen if nuclear weapons were ever again used. These uncertainties, combined with the catastrophic results sure to follow from a maximum escalation of the exchange, are an essential element of deterrence.
>
> My own view remains that a full-scale thermonuclear exchange would constitute an unprecedented disaster for the Soviet Union and for the United States. And I am not at all persuaded that what started as a demonstration, or even a tightly controlled use of the strategic forces for larger purposes, could be kept from escalating to a full-scale thermonuclear exchange.

The new secretary of state, Alexander Haig, has echoed these views.

It is common practice in the Pentagon to carry out simulated war games in order to develop an improved understanding of the wartime decision-making process. In a typical mock war in Europe, one group of participants, senior American officers and government officials, took the part of U.S.–NATO decision makers (the blue team) and another the Soviet-Warsaw Pact (the red team). The game started out with a Soviet aggression using conventional weapons, but at one point when the West appeared to be losing, the blue team called for the use of a single nuclear explosion on a Soviet tank force to demonstrate the West's seriousness and to force the Soviets to halt their aggression. In order to show they were not to be intimidated, the red team then dropped five A-bombs on NATO airfields, destroying all the planes and equipment on the ground. Blue responded with twenty bombs on Soviet staging and supply centers. And so it went, up and up. The result was the devastation of not only Europe but the United States and the Soviet Union as well. The lesson of that game: No one knows how to stop a nuclear war because no one knows how to lose.

Although such war games are not the same as an actual war because the players know they are not dealing with real human lives and atomic bombs, they demonstrate the dangers of even a so-called limited nuclear war. Even if someone were confident that the world powers could stop short of the ultimate insanity of all-out nuclear war, the risks that they might fail to control escalation are so great that we can never afford to put it to the test. Avoidance of a nuclear conflict of any kind or size by strengthening the inhibitions against the first use of nuclear weapons is a key objective of national security policy. This must be kept in mind as decisions are made to procure new weapons such as the MX. Regrettably every day we hear more and more talk of strategies and weapons to fight and win a nuclear war.

Defense Against Atomic Attack

A final reality of nuclear war is that there is no way to protect the population from an atomic attack, nor is there likely to be in the future. This is true of not only active defenses designed to prevent missile warheads and bombers from hitting cities or industrial targets but also of passive or civilian defense designed to reduce damage and casualities and to permit recovery in its aftermath. The view that if we only had missile defenses, antiaircraft forces, and civil defense shelters and evacuation plans, we could survive and win is a delusion, spread to make plans for a nuclear war more acceptable. Although by tradition the Soviet Union is strongly defense oriented, its programs are often exaggerated in the United States to make it look as if their leaders accepted this myth. No Soviet generals could safely admit that they would lose a nuclear war, but the political leaders over and over again have emphasized what a disaster a nuclear conflict would be for all concerned.

The difficulty in trying to develop active defenses against an atomic attack is the fundamental nature of a nuclear explosion.

Even a single explosion can cause such catastrophic damage that one alone would be an unparalleled disaster for the victim nation. Yet both the Soviet Union and the United States have many thousands of strategic nuclear warheads with the explosive force three to a thousand times as great as the bombs that destroyed Hiroshima and Nagasaki. Most defenses are considered good when they succeed in destroying 10 percent of the attacking force. However, even 90 percent effectiveness would not reduce the casualties and damage from a nuclear attack to acceptable limits. A 99 percent or even higher efficiency would be needed to save our nation. This has never been feasible in the past, and it does not appear so in the foreseeable future even with advanced technologies.

The offense has a tremendous advantage over the defense since it can determine the timing and scale of the attack and select the area to be hit. Even if the offense chose to attack only lightly defended, less critical targets, the devastation would be tremendous and the radioactive contamination widespread. On the other hand the concentration of many warheads on a single important and highly defended target can overwhelm the defenses, saturate them, and literally run them out of ammunition. It would not be much satisfaction to the few survivors in New York, Chicago, or Los Angeles to be reassured by the Defense Department after their areas were destroyed that the Soviets had had to expend more nuclear warheads than we had used in their defense. Yet this is the kind of logic that is used by those who argue we should rely on defense to protect us against a nuclear attack.

Nor is civil defense the answer to survival in a nuclear war. Possibly it could somewhat reduce the casualties, but the cumulative effects of such a war would be so devastating that civil defense would also be overwhelmed. As early as twenty-five years ago when the atomic stockpiles were a fraction of today's, Val Peterson, U.S. civil defense administrator, said, "If the whole 170 million Americans have air raid shelters, at least 50 percent of them would die in a surprise enemy attack. In the last analysis, there is no such thing as a nation being prepared for a thermonuclear war." Even with extensive advance planning, there would be few medical facilities to care for the millions of injured, no transportation to bring in aid from the outside, and little food to keep the survivors alive. Evacuation outside cities is of little help when tens of thousands of square miles are contaminated by radioactive fallout. Civil defense cannot make nuclear war acceptable.

Only one conclusion can be drawn from these fundamental realities: Nuclear war must be avoided if our civilization as we know it today is to survive. Bernard Brodie, the first nuclear strategist, recognized this in 1946, only a year after Hiroshima and Nagasaki, when he wrote, "Thus far, the chief purpose of our military establishment has been to win wars. From now on its chief purpose must be to avert them."

6 Prevention, Yes; Nuclear War Fighting, No

Mutual Deterrence

Our political and military leaders did not quickly absorb Brodie's dictum. Not until the late 1950s did they face up to the reality that the John Foster Dulles policy of massive retaliation with nuclear weapons was no longer a credible basis for maintaining our security. The consequences of even a small nuclear conflict combined with the absence of any hope of attaining defenses that could realistically reduce these to acceptable levels made the threat of nuclear retaliation against all types of aggression both unbelievable and dangerous. Certainly by the time of the Cuban missile crisis in 1962, and probably even earlier, both the United States and the Soviet Union had sufficiently large and effective nuclear forces to be able to wreak irreparable damage to the other nation. Nuclear weapons realistically had to be relegated to the function of deterring an atomic strike or aggression on such a large scale that the attacker would fear that escalation to the nuclear level would be almost inevitable. Thus as the nuclear stockpiles of the United States and the Soviet Union grew, the current strategic policy of deterrence through fear of retaliation gradually evolved.

Since the prevention of a nuclear war became the primary security objective, deterrence had to be two-sided or mutual in order to ensure that neither American nor Soviet leaders would have any incentive to launch a nuclear strike. Thus paradoxically it was in the U.S. interest for the Soviets also to have an invulnerable strategic deterrent force because this would reduce incentives for them to unleash their nuclear weapons in a time of crisis. If they worried that we might strike first at their vulnerable forces, they might plan even in peacetime to launch rapidly their missiles as soon as their radars and computers warned of an attack. This tactic raises the specter of an accidental nuclear war. Furthermore if there is a situation where they see we could gain great advantage by a first strike, they might decide to launch a preemptive strike in a period of high tension. Since both these alternatives could lead to disastrous consequences for the United States, it was and still is in our interest for the Soviets to be confident that their strategic forces can survive a U.S. first strike.

This policy of Mutual Assured Destruction in which both sides can retaliate so as to produce unacceptable damage to the other has frequently been referred to by the disparaging acronym MAD because it purportedly relies on destroying cities and killing people. Detractors claimed such retaliation to be immoral and incredible. However, U.S. war plans never called for targeting cities or people per se; this is a blatant misrepresentation of the actual policy. They have always called for aiming at military targets or industrial installations that supported military or recovery operations. Secretary Brown in January 1981 reaffirmed that "the United States has never—at least since nuclear weapons were available in significant numbers—had a strategic doctrine based simply and solely on reflexive massive attacks on Soviet cities and populations." Nevertheless in the course of destroying the Soviet military establishment, cities would also be devastated and millions killed.

Deterrence can be obtained with a finite limit on strategic force levels. Under this doctrine, sufficiency is a meaningful criterion. On the other hand if war fighting is the strategy, there is no upper limit; we will never have enough weapons. Mutual deterrence reduces pressures for an arms race; war fighting drives it ever faster.

Recently the flexibility of our strategic targeting and command and control has been improved to provide options for limited responses as well as large-scale ones. This is not bad since we should never be left with world disaster as our only choice. Now our deterrent is believed more credible because the retaliation can be tailored to the seriousness of the aggression. It should not be forgotten, however, that a small nuclear response could escalate uncontrollably to a major conflict. Moreover improved flexibility does not mean that we must be able to threaten the Soviet deterrent or even a major component of it, such as their ICBMs. We lack the forces to do that, but the MX would change that picture.

For our deterrent force to be credible, it must not only be able to launch our weapons toward the Soviet Union but it must also be able to ensure its warheads reaching and destroying their targets. Truly effective defenses could undermine mutual deterrence, and even partially effective ones could reduce confidence in it. In the 1960s the Soviets developed and began deploying around Moscow ballistic missile defenses (ABMs). Even though we did not believe these were very effective, the deployment caused considerable concern in American security circles, and many weapons programs were undertaken to ensure that Soviet missile defenses could be penetrated or, even better, overwhelmed. Multiple warheads and later MIRVs were put on each missile so that potential Soviet defenses would be saturated—literally have their interceptor stockpile exhausted. Although an effective counter to ABMs, MIRVs had a disastrous effect on the mutual strategic deterrent balance.

Coincident with this approach to cope with ABMs by procuring new weapons, discussions were started with the Soviet Union to control the size of the ABM threat through arms control. These negotiations culminated in the ABM Treaty of 1972 in which both the United States and the Soviet Union agreed to limit their ballistic missile defenses to such low levels that neither country could be worried that they would seriously degrade either nation's retaliatory capabilities or undermine the mutual deterrent.

To achieve this critical objective of avoiding the outbreak of nuclear war, our military programs must be continually monitored to ensure that they not only do not undermine the deterrent, as do ABMs, but also that they do not create, even inadvertently, any incentive for a Soviet first strike. In January 1980 Secretary of Defense Harold Brown stated that "crisis stability means insuring that even in a prolonged and intense confrontation, the Soviet Union would have no incentive to initiate an exchange and also that we would feel ourselves under no pressure to do so." This is not a new concept. In 1971 President Nixon and his secretary of defense, Melvin Laird, categorically affirmed that the United States was not acquiring a first-strike capability and would not take any steps that the Russians could even construe as moving toward such a position. The Defense Department programs have not always been consistent with such stated principles, however. Secretary Laird strongly supported MIRV programs, which eventually led to a countersilo threat, and Secretary Brown has pushed for the MX with its silo-busting characteristics, useful only in a first strike.

The lure of fighting a nuclear war never seems to go away. Analysts in research institutes, often in conjunction with designers of weapons systems, try to find ways in which nuclear weapons can be made militarily useful. The concept of spending billions on the most efficiently destructive weapons that will never be used in an actual conflict is counter to conventional military planning. Military men are taught to fight, and deterrence is inconsistent with this indoctrination. The horrors of nuclear war seem unreal and beyond human conception and so are given little attention in day-to-day activities.

Former Secretary of Defense James Schlesinger, Laird's successor, came to the job with a background in military systems analysis. In 1973 he reversed the Nixon-Laird policy of not procuring weapons that could threaten Soviet missiles and on July 1, 1975, even threatened the Soviets with a first strike on their strategic forces. He started weapons programs specifically designed to endanger Soviet ICBMs. MIRVs with higher accuracy and explosive force were approved. This critical about-face in strategic policy was largely ignored, and perhaps even unrecognized, by President Nixon and his successor, President Ford. There is no public indication of a thoughtful review of this new policy at the presidential level.

On September 11, 1974, in testimony before the Senate Foreign Relations Committee, Secretary Schlesinger made a major ploy to make nuclear war seem more acceptable. He lent his authority to a scenario in which the Soviets would launch a first strike against our 1,054 ICBM silos using 1 mt warheads. He claimed that this attack might destroy 90 percent of our ICBMs and that a U.S. president might not retaliate after such an attack for fear that the Soviets would then destroy our cities and kill 95 million Americans. This public statement seems curious, for the secretary of defense knew that our security depends on the threat of our retaliation to deter Soviet aggression. The secretary buttressed this view with calculations that only 800,000 Americans would be killed by a Soviet attack designed to knock out our ICBMs and that the casualties could be even less with proper civil defense.

The senators, astounded by this attempt to make nuclear war seem so benign, commissioned the Office of Technology Assessment to do an independent study of the collateral effects of a successful attack on our Minuteman ICBMs. This study showed that a more realistic estimate of the number of American casualties could be 20 million, and the Defense Department later concurred in the revised estimates. This attempt to make nuclear war acceptable failed, but the concepts linger on and indeed have been given new life.

The Schlesinger scenario, albeit with much larger American casualties, is still used by those denigrating our deterrent and exaggerating Soviet threats. It is on such unrealistic scenarios upon which are based the fears that our existing Minuteman ICBMs will be vulnerable and therefore must be replaced by the new MX missile in multiple protective shelters.

This scenario ignores the realities of nuclear war. In his annual report for fiscal 1979 of February 2, 1978, Secretary of Defense Harold Brown cogently explained why the vulnerability of the current Minuteman missiles "would not be synonymous with the vulnerability of the United States, or even of the strategic deterrent":

> In recognizing that the MINUTEMAN vulnerability problem is a serious concern for us, we also realize that the Soviets would face great uncertainties in assessing whether they would have the capability we fear—and still greater uncertainties as to its military or political utility.
>
> On all the technical judgments—how accurate the missiles are, how reliable, how well the system would work in actual practice, whether they could explode two reentry vehicles on each silo without excessive fratricide, or only one—we, quite properly, are conservative, from our point of view.
>
> Similarly, the Soviets must make cautious assumptions from their perspective. In particular, they must recognize the formidable task of actually executing (as planned) a highly complex massive attack in a single cosmic throw of the dice.
>
> Even if such an attack worked exactly as predicted, the Soviets would face great risks and uncertainties.
>
> First, they would necessarily have to consider whether the US missiles would still be in their silos when the attack arrived, or whether, given our capability to have unambiguous confirmation of a massive attack, we would launch from under the attack.
>
> Second, and more important, an attack intended to US silos could kill at least several million Americans and would leave untouched at least the alert bombers and at-sea SSBNs with thousands of warheads. The Soviets might—and should—fear that, in response, we would retaliate with a massive attack on Soviet cities and industry. The alleged "irrationality" of such a response from a detached perspective would be no consolation in retrospect and would not necessarily be in advance an absolute guarantee that we would not so respond.
>
> In any event, any Soviet planner considering US options would know that, besides massive retaliation, the surviving US forces would also be capable of a broad variety of controlled responses aimed at military and civilian targets and proportioned to the scale and significance of the provocation.

Thus the Carter administration's principal spokesman argued that the realism of the Soviet threat to our ICBMs was open to serious question. Although the Soviet forces will be greater in the 1980s than when Brown made that statement, its logic will be equally applicable then.

Nevertheless the concept of fighting a nuclear war gains more supporters every day. The Carter White House, feeling political pressure from the cold war warriors and Republicans who believed a nuclear war could be won, leaked to reporters in early August 1980 that it had adopted a new strategy for fighting limited nuclear wars. According to Presidential Directive 59 (PD 59), signed July 25, 1980, we would now prepare to fight a long but limited conflict by using precise nuclear strikes against military targets not cities. The MX missiles, because of their pinpoint accuracy, would be the linchpin of such a strategy; justification of the cost of these missiles was in fact one reason for promulgating it. This was quite a change from President Carter's and Secretary Brown's previous positions, for they had repeated that there was no known way of keeping a nuclear conflict small. Even six months after the directive was issued, Secretary Brown reiterated that "a 'limited' exchange is unlikely to remain limited."

PD 59, however, gave presidential authority to and perpetuated the risky concept that precision nuclear weapons will enable us to fight a nuclear war and survive. This is a dangerously misleading concept, for casualties cannot be kept to manageable levels. There is no such thing as a sanitized nuclear war. The new precision weapons, including the advanced MX, will permit a warhead to explode half of the time within three hundred feet of the aim point instead of the 1,200 feet for our first-generation Minuteman III MIRV warheads. But the yield or explosive force of each of these earlier Minuteman warheads was only 150 kt, while that for the MX warheads will be 335 kt and perhaps eventually even larger. The Minuteman warheads would have destroyed homes about 2½ miles away from the point of detonation, while the MX will extend this damage radius to about 3½ miles, a 40 percent increase. Thus although the MX will hit closer to its aim point, the radius of the area within which there will be widespread casualties and collateral damage to civilian housing will extend far beyond that of the warheads it replaced. The area will be increased two and a quarter times.

Furthermore in order to destroy the hardened targets such as missile silos or shelters for command authorities, two targets specifically singled out in the new strategy, the warheads would have to explode close to the surface and thus produce large areas of lethal radioactive contamination. Such weapons are hardly ideal to make nuclear war safe for the citizen.

This new strategy for fighting limited wars calls for targeting the hardened bomb shelters, which protect Soviet officials in control of their atomic forces. It is mystifying how we are to keep this "sanitized" nuclear war limited after we have killed all the Russian leaders. With whom will we deal if it is to be limited or ended? Why will those still surviving play the war game by our rules and not respond with all-out strikes? Who will be there to say "We quit," if they want to surrender?

Fighting a nuclear war will lead to disaster for the human race. No one can be sure of keeping it limited once it starts; escalation to an all-out conflict in which thousands of warheads are used on all types of targets is almost certain. As former national security adviser, McGeorge Bundy, wrote in October 1969, "a decision that would bring even one hydrogen bomb on one city of one's own country would be recognized in advance as a catastrophic blunder; ten bombs on ten cities would be a disaster beyond history; and a hundred bombs on a hundred cities are unthinkable."

7 The Growth of Strategic Forces

Over the years strategic policies and the forces to support them have evolved as new weapons or new theories were developed. In some cases the weapons were procured in order to satisfy certain policy needs; yet far too frequently the weapons were developed and procured and then the policy modified to justify them. To some extent Presidential Directive 59 and the MX are an example of the latter case.

In the post–World War II years the United States produced large numbers of intercontinental bombers; the World War II B-29s were supplanted with B-36s, armed first with fission weapons and later hydrogen bombs. These planes were designed to give muscle to the policy of massive retaliation in the event of Soviet aggression anywhere in the world. The Soviet Union, starting considerably behind the United States, attempted to build a long-range bomber force of its own by procuring Bison jet and Bear turboprop bombers with intercontinental range. It also put a tremendous effort into building antiaircraft defenses in order to protect its cities from our bomber threat. In the mid-1950s the United States entered into a program for the procurement of more advanced B-47 bombers and then later the B-52s in order to eliminate the feared bomber gap from the Soviet buildup.

The Soviet Union, however, opted out of the race after building only about 200 Bison and Bear long-range bombers and instead turned its attention to ballistic missiles. Their initial ICBM, the very large and cumbersome SS-6, was first tested successsfully in August 1957 and in the next two years went through a thorough testing program so that it could have been operational by 1960, or perhaps even earlier. This same vehicle was used to launch the first Sputnik satellite on October 4, 1957, more than a year before the United States was ready to do so, and this caused great concern in American security circles.

Although ballistic missiles were given highest priority in the United States, there was no way that we could be in a position to deploy our first ICBMs by the time that the Soviets had demonstrated that the SS-6 could be ready for use (1959 or 1960). In view of the extensive Soviet testing program, which demonstrated the SS-6 to be a reliable missile, it was logically assumed that they were proceeding to deploy this ICBM, though it was not possible for us to locate any deployment sites. Our U-2 reconnaissance plane was capable of only limited photographic coverage of the Soviet Union, so the operational status of this missile could not be readily determined. Air force intelligence postulated widespread ICBM deployment by 1960, while the navy estimated the number of operational missiles to be very low or even nonexistent. CIA estimates were somewhere in between, but there were few people who did not subscribe to the existence of some missile gap even though its size and significance was uncertain. Fears of a Soviet surprise attack mounted as the decade ended. President Eisenhower, however, did not believe that the Soviet force was sufficient to threaten our very large force of bombers, many of which were kept on an alert status to avoid being caught by surprise on the ground.

Late in 1960, some months after the U-2 was shot down over the Soviet Union, the United States acquired the ability to photograph the Soviet Union from satellites. The first useful pictures were taken in November 1960, and gradually over the next six to nine months pictures of the entire Soviet Union were obtained. This new wide area coverage gradually showed during 1961 that the Soviet Union had never gone forward with the deployment of its first ICBM, probably because it was too big and unmanageable. At the most only a handful were deployed at Plesetsk, a site northwest of Moscow, which was also used for certain operational testing and space flights. Thus by the end of 1961 the missile gap had been imploded, and instead, because of the high priority given to the first-generation U.S. Atlas and Titan and the second-generation Minuteman ICBM and Polaris submarine missile programs, the Soviet Union ended up way behind.

Beginning in the early 1960s, the United States went rapidly ahead with a broad but orderly program to build up a balanced strategic deterrent force. Survivability was recognized as the most critical characteristic for such a force, and the vulnerable Atlas and Titan Is were replaced with a thousand Minuteman and fifty-four Titan II missiles in blast-hardened underground silos. The United States also built forty-one nuclear-powered submarines, each armed with sixteen strategic ballistic missiles; these are not considered to be vulnerable for the foreseeable future. The United States completed the procurement of about six hundred B-52 intercontinental bombers to provide the third leg of this strategic triad. By reliance on this three-pronged triad of strategic delivery vehicles, the United States was protected against any unforeseen technological or operational breakthroughs that could make the deterrent as a whole vulnerable.

Each part of the triad had strengths and weaknesses. The land-based ICBMs had the best command and control because of the simplicity of communications between the highest authority and the officers with access to the firing circuits. High accuracy is most easily achieved with fixed, land-based ICBMs. They can reach their targets within thirty minutes of launch and so provide for rapid response if required. But once they are out of their silos, they cannot be recalled, so a nuclear war could not be avoided. Because of their known, fixed location, they can be the target of accurate missile warheads.

The bombers have a slower reaction time (six to eight hours to reach their targets), but they do not have to be committed irrevocably to a nuclear attack when they take off from their airfields. Inflight refueling enables them to remain airborne for a long time, and until they cross the borders of the Soviet Union, they can always be recalled. Although they are quite vulnerable on the ground to a nuclear explosion, one-third of the force can be kept on fifteen-minute alert to prevent their being destroyed in a surprise ICBM attack. Once in the air, they are essentially invulnerable until they have to penetrate Soviet defenses. The extensive Soviet antiaircraft defenses could in the past be overcome by flying close to the ground to avoid radars and by bombing a path through the ground defenses. Since airborne radars for tracking low-flying planes are now greatly improved, it is planned in the future for the bombers to stay outside the perimeter of enemy defenses and to launch attacks through them with large numbers of small, low-flying, unmanned cruise missiles.

The submarine ballistic missile force is by far the most survivable of all elements of the triad, since there are no developments on the horizon that could make them vulnerable to a Soviet first strike. Although a single submarine might be sunk, there appears no way in which a fleet of about thirty submarines that would be at sea at any one time can be destroyed on short notice. The greatest disadvantage of the submarine force is the problem of maintaining two-way communications without disclosing the location of the ship. This difficulty creates problems for command and control. In order to remain undetected, submarines must stay below the surface of the water, which screens out all but extremely low-frequency radio waves. A number of redundant systems have been developed to provide safe communication, but none of these are ideal for use in very deep water.

Difficulties in knowing the precise location of the submarine and its relative motion at time of launch also make it harder for ballistic missiles launched from a submarine to be as accurate as those launched from a fixed presurveyed point on land. This is critical only if the warhead is aimed at a very blast-resistant target. Submarine missiles can easily destroy most other military targets. In the future when guidance corrections can be given to a warhead after it is well on its way to its target, even very hard targets can be destroyed by submarine missiles. The overall costs of a ballistic missile launched from a nuclear-powered submarine are high, but this is compensated for by its being virtually invulnerable to attack and even to being located.

By 1966 this well-balanced U.S. strategic deterrent force, consisting of 1,054 ICBMs, 656 submarine-launched ballistic missiles (SLBMs), and approximately 600 long-range B-52 bombers, was completely operational. The Soviets in the meantime were far behind. Their missile submarines were just beginning to become available, and their second-generation ICBMs were deployed in relatively limited numbers. Their 200 Bison and Bear intercontinental bombers had not been significantly improved since they were first deployed in the mid-1950s. In the late 1960s, the Russians struggled to catch up.

On the other hand, while the Soviet Union had lagged in offensive strategic systems, it had continued to concentrate on its defensive arsenal. Generation after generation of new radars dotted the Russian landscape. The 3,600 surface-to-air missiles deployed around Moscow in 1955 were replaced only a few years later by more modern defensive missiles, which were deployed in vast quantities throughout the Soviet Union. Interceptor aircraft were procured in large numbers. No expense was spared in order to satisfy the strong Soviet interest in defense.

Beginning in the late 1950s the Soviet Union began an extensive development and test program for ballistic missile defenses. When the United States was still in the stage of preliminary studies, the Soviet Union was testing concepts using intermediate-range ballistic missiles as their targets at their antiballistic missile (ABM) test site at Sary Shagan, near Lake Baikal in the southern Soviet Union. In the mid-1960s they started deployment of an ABM system around Moscow, and it was anticipated that defenses would be built around other Soviet cities. Satellite photographs picked up evidence of new construction in many places, but as the construction continued and intelligence became more extensive, it turned out that none of these other sites were ballistic missile defense installations. The only actual Soviet ABM deployment was around Moscow and about 1968 the construction on even this system was slowed to a crawl. About this time the Soviet Union first expressed an interest in the U.S. proposal to negotiate an arms control agreement to limit ABMs. Previously all Soviet officials and scientists had supported ballistic missile defenses and argued that only offensive systems should be limited. When the SALT I talks actually got underway in 1969, it was the Soviets who were soon pushing primarily for an ABM Treaty, while now the United States was insisting on offensive systems being controlled as well. This was not the first or the last time that the two countries had exchanged their positions on an arms control issue.

In the 1960s the United States had put off any decision to deploy ballistic missile defenses. All of the U.S. designs, and the Soviet ones as well, appeared easy to penetrate and overcome. Although there were many debates on the virtues of various systems, Secretary of Defense McNamara managed to forestall any commitment of funds for deployment of ABMs. However, in September 1967 under strong political pressure, President Johnson authorized the deployment of the so-called Sentinel ABM, a limited but nationwide system designed to cope with an anticipated light Chinese ICBM threat. ABM enthusiasts viewed this as the opening wedge toward a full-scale anti-Soviet ABM, while skeptics accepted this as the best way of relieving political pressures until such time as an agreement could be reached with the Soviet Union to limit strategic arms. In 1969 the new Nixon administration scrapped the Sentinel anti-Chinese version, but after a strenuous public debate authorized the Safeguard ABM designed to protect missile sites, not cities. Again the proponents welcomed and opponents decried this as an opening wedge, since they did not believe that approval and funds could be obtained for a full-scale system.

Although the United States did not believe that ABMs could be made effective, it did carry out extensive development programs with its offensive missiles to ensure that any Soviet ABM system could be overcome and that our deterrent would not be degraded. The most important of these programs was the development and eventual deployment of MIRVs. The basic principle was to make the final stage of the missile, which contained in the U.S. case from three to fourteen warheads, a sort of "bus," which successively released its warheads on trajectories toward separate targets. The bus would be aimed on a path to a first target, drop off its warhead, then move to another trajectory toward a second target, and then unload a second warhead. This process was repeated until all warheads had been sent on their trajectories toward their separate targets. With large numbers of warheads on a single missile, it was possible to overwhelm an ABM system and guarantee that a warhead would reach the specific target.

Although this principle was ideal for ensuring that an ABM system would not undermine the deterrent, it had a corollary characteristic that is now profoundly destabilizing the arms race. Since each missile could fire several warheads at different targets, in theory a single missile with MIRVs could threaten several missiles of the other side. As long as there was only one warhead on each missile, there would never be any advantage for a missile to attack another. The exchange ratio between missiles destroyed and missiles used in an attack would always be less than one, an unfavorable position for the attacker because no missile with a single warhead could have complete assurance of destroying another missile in its silo. When a missile carries several warheads, however, this disadvantage for the attacker in theory could be reversed and turned into a gain. The exchange ratio could be greater than one, and thus favorable to the attacker, if the accuracy and reliability were sufficient to give each warhead a high probability of destroying a missile silo. Thus now there could be a real gain by launching a first strike against the opposing side's land-based ICBM force. MIRVs now created incentives to initiate a nuclear strike; nuclear war fighting would now seem a more attractive tactic; and the dangers of a nuclear war would now be radically increased.

The U.S. first tested MIRVs in August 1968 and then went ahead with high-priority programs for their production for both the Minuteman III ICBM and the Poseidon submarine missile. Deployment of these new MIRVed missiles began in 1970 with three warheads on the Minuteman III and up to fourteen on the Poseidon (the average number of Poseidon warheads is reported to be ten). At this time a major debate arose in U.S. security circles over whether to seek an agreement to limit MIRVed missiles in SALT I before either side had deployed them. Rather than negotiate, however, the United States went ahead with its deployment programs. The Soviet Union, running about five years behind the United States in MIRV technology, tested its first MIRVed ICBM in the summer of 1973. Deployment of their ICBMs with MIRVs did not begin until late 1975. Submarine-launched MIRVed missiles were even further behind; they began to be extensively installed on Soviet submarines only in 1979. But by 1980 it was obvious that Soviet MIRVed missiles would soon threaten our land-based ICBMs.

By the end of 1979 when the MX decision was made by the Carter administration, both the United States and the Soviet Union had large numbers of nuclear weapons with strategic missiles and bombers for delivering them. The numbers of the key weapon types in their respective strategic arsenals in Jan. 1981 are summarized in table 7.1. Since the dynamics of an arms race is important in evaluating the situation, the levels that existed in mid-1972 when the SALT I agreements were signed are included for comparison.

Our technical intelligence capabilities, particularly observation satellites to photograph the location of deployed systems, pro-

Table 7.1
U.S.–U.S.S.R. strategic force levels

					1986			
	Mid-1972		Jan. 1981		With SALT II		Without SALT II	
Weapon type	U.S.	U.S.S.R.	U.S.	U.S.S.R.	U.S.	U.S.S.R.	U.S.	U.S.S.R.
Total delivery vehicles	2,167	2,110	1,975	2,504	2,088	2,250	2,112	2,770
Total ICBM	1,054	1,533	1,052	1,398	1,052	1,200	1,052	1,604
MIRVed ICBMs	139	0	550	652	550	820	550	1,190
ICBM warheads	1,332	1,533	2,152	5,354	2,152	6,080	2,152	9,110
Total SLBMs	656	437	576	950	640	950	712	1,016
MIRVed SLBMs	160	0	496	192	640	380	664	444
SLBM warheads	2,096	437	4,656	1,334	6,344	2,470	6,584	3,740
Intercontinental bombers	457	156	348	156	348	100	348	150
Strategic cruise missiles	0	0	0	0	2,600	0	3,400	100
Total MIRVed delivery vehicles[a]	269	0	1,046	844	1,320	1,200	1,384	1,734
Total force loadings (warheads and bombs)	5,598	2,282	9,000	7,000	13,100	8,750	13,260	13,150

Source: Mid-1972 and Jan 1981, U.S. Department of Defense, *Annual Reports.*
[a]Includes bombs configured to launch cruise missiles.

vide high confidence in the reliability of our estimates of Soviet strategic weapons, past and current. Construction of new missile launch sites normally can be detected two or more years before they become operational. Submarine construction and fitting out can be watched. The characteristics of new missiles can be determined through observations from land, sea, air, and space of the test firings of both land- and sea-based types. Some Soviet missiles have been tested in both single-warhead and MIRV versions, so it has not always been possible to know precisely from unclassified data the exact number of each type actually deployed. Nevertheless any errors are not very significant. The SALT II Treaty counts all such missiles as MIRVed with the maximum number of warheads with which it has ever been tested. Intercontinental bombers can be spotted on airfields and cannot be concealed. As part of the SALT II Treaty, the Soviet Union was required to declare numbers of each type of strategic delivery vehicle, and these declarations served as additional confirmation that our intelligence in the past was highly accurate.

Estimates of future deployments are, of course, much less reliable. Since new systems cannot be deployed overnight and all require extensive testing, estimates for a few years in advance are quite reliable. But predictions for five or more years in the future are subject to increasing uncertainty. For example, when a given type of missile is in the process of being deployed, there is no way of telling whether a total of 200 or 400 or 600 will actually be placed in the field. All that can be done is to note when new starts on site construction stop. Then it may be concluded that two or three years hence, all weapons of that type will have been deployed and the total number will be known. As long as the program is continuing, it is impossible to predict confidently when it will end. This difficulty led to some underestimates of Soviet ICBMs in the late 1960s, about the only time the force levels have been underestimated.

It is likely that deployments of future types of weapons may be much more difficult for intelligence to monitor. For example, were the Soviets to develop mobile systems similar to the MX, the actual number deployed might be difficult to determine. The United States is setting a precedent in its deployment of mobile ICBMs with its MX. We have designed the system with a view to making it possible to verify the actual number deployed; however, we cannot be certain whether the Soviets will follow this precedent, particularly since their liquid-fueled missile technology is not suited to a deployment scheme similar to that proposed by the Carter administration for the MX. In addition, cruise missiles, with which the United States is again taking the lead, are small and relatively easy to hide. Once these systems are fully developed and ready to be deployed on land and sea, it will be extremely difficult to keep track of the exact number that are operational. For the moment the Soviets are not ready to deploy long-range cruise missiles, but when inevitably they will, we may regret having started an uncontrolled race in which we must run blindfolded.

The SALT II Treaty would have put some constraints on these future deployments and would have made it much easier to verify the extent of the Soviet strategic threat. Table 7.1 includes estimates of the force levels that might be available to both countries in the mid-1980s were SALT II to be ratified. These numbers are reasonably accurate, although to some extent they will depend on the choices that each country makes within the constraints of the SALT II Treaty.

Estimates of the force levels in the absence of SALT II are also included in table 7.1. These are much less certain, but even here their reliability is fairly good because it might be later than 1986 before the full effect of an uncontrolled arms race would be evident in the numbers of weapons deployed. A considerable lag time exists between the decision to procure new weapons and their actual deployment. This lag would be reflected more in the

U.S. force levels than those in the Soviet Union since, at the moment, the only U.S. active production lines are for the Trident submarines and the Trident I missile. The Soviet Union, on the other hand, is in the process of replacing some of its older weapons with new models and could accelerate any buildup. Thus the prediction for the U.S. levels have only a small uncertainty, while the Soviet levels could be significantly higher if they made the decision to continue their programs at a rapid rate and did not scrap their older models. Extrapolation beyond the mid-1980s becomes increasingly unreliable and will depend greatly on the international political climate in the next few years.

8 U.S.–U.S.S.R. Strategic Balance

Public discussions on national security, military budgets, and arms control have been largely focused on the numerical balance between the strategic forces of the United States and the Soviet Union. Only a few relatively simplistic qualitative measures, such as missile size, have received any attention. Strategic doctrine and policies have been considered too esoteric for the average citizen. The tendency to focus attention on a few numbers such as some of those in table 7.1 has often been used to mislead the casual observer.

Because both countries now have thousands of nuclear weapons and will probably continue to have more than enough survivable ones to devastate each other under any circumstances, meaningful military superiority is not a practical goal. The paramount strategic problem is how to prevent such weapons from being used. Nevertheless, the questions of strategic inferiority, parity, or superiority have become hot political issues, which frequently override military considerations in such decisions as the procurement of weapons like the MX missile. Opinions on the state of the strategic balance range from the view that the Soviet Union already has an overwhelming advantage to the view that the United States has and can maintain superiority indefinitely. Still others look at the question as irrelevant because of the large numbers of weapons on each side. With asymmetrical forces and geopolitical situations between the United States and the Soviet Union, it is easy to select criteria to support all points of view.

A number of questions should be addressed to the issue of the strategic balance. Does either nation have today or could it have in the foreseeable future the strategic forces that can be used to gain some specific military advantage? Is a disarming first strike a strategy that must be realistically reckoned with? Is a more limited strategic attack ever a militarily or politically rational action? Is the balance such that there can develop dangerous incentives for initiating a nuclear conflict? How fragile or sensitive is the strategic balance to the development of new weapons systems?

The strategic balance must not only be evaluated in strictly military terms but must also be analyzed in light of the perceptions of that balance by the U.S. public, by Soviet leaders, by our allies, and by the rest of the world. Can a perceived imbalance be translated into political advantage or disadvantage even without any military action occurring? What will be the actual consequences of misperceptions, which are sometimes fostered to promote military programs, undercut arms control, or further political objectives? Unfortunately perceptions are subjective and not necessarily based on facts, so they can be and often are used to support some specific point of view. Still more unfortunately, our security suffers when this is done.

Disarming First Strike

The critical function of a strategic force is to provide deterrence, to have the ability to survive any potential first strike and still be able to retaliate, producing unacceptable damage to the aggressor's military establishment and the industry that supports it. If either nation could seriously contemplate being able to launch a disarming first strike and survive virtually unscathed, then deterrence would be undermined and the security of the other nation would be in jeopardy. Such a disarming first strike would produce the ultimate military advantage. Fortunately for the peace of the world, this type of imbalance of forces is not likely either today or in the foreseeable future. Neither nation now has the weapons that can threaten a significant part of the other's deterrent. In the future a single component could appear threatened but not the entire deterrent force.

Survivability is the key characteristic of a strategic force that determines whether a disarming first strike is feasible. In this area the Soviet strategic deterrent as a whole is far inferior to that of the United States. Beginning in 1960 the United States quite wisely, but not necessarily intentionally, decided to rely on a triad of submarine-launched ballistic missiles (SLBMs) supported by land-based ICBMs in hardened silos and intercontinental bombers. This reliance on a triad means that our deterrent as a whole is not becoming vulnerable as ICBMs in fixed silos become less survivable a few years from now. On the other hand the Soviet Union depends primarily on fixed land-based ICBMs (75 percent of its strategic forces). The submarine and bomber legs of their deterrent are much less survivable than those of the United States.

Despite the Soviet inferiority in survivability, the United States is far from having a disarming first-strike capability. The Soviet Union is even less able to carry out such an attack. A fundamental feature of the strategic balance is that with or without a SALT agreement, neither side will be able to mount a successful disarming first strike today, in this decade or in this century.

First let us look at the United States. We now have thirty-six large nuclear-powered submarines equipped with about 5,000 strategic warheads. The number of submarines is dropping as we phase out the old, first-generation Polaris ships with single warhead missiles and replace them slowly with very large Trident submarines with longer-range Trident I missiles. Although submarine-launched warheads do not now have the accuracy and yield combination to threaten extremely hard targets, such as hardened ICBM silos, they are quite capable of destroying most military installations in addition to producing vast devastation to Soviet industry and population centers. Approximately 60 percent of our submarines would be at sea at any one time and therefore be invulnerable. There is no foreseeable antisubmarine warfare (ASW) threat to these ships, and in fact their invulnerability probably will grow since many of even the older ones are now being reequipped with Trident I missiles, which have a range of 4,000 nautical miles as compared to 2,500 earlier. This range would permit them to launch their missiles as soon as they leave U.S. ports, while stationed in protected coastal waters, and when cruising in millions of square miles of ocean. No technological developments currently on the horizon could make our missile submarines become vulnerable. In a speech at the Naval War College on August 20, 1980, Secretary Brown admitted that future threats to our submarine force were "hypothetical," discrediting certain Defense spokesmen who had been attempting to build up Soviet antisubmarine warfare in order to justify the MX land-based missile program.

This submarine missile force is backed up by a large number of intercontinental bombers, primarily B-52s, which by 1985 will be equipped with several thousand long-range strategic cruise missiles. These can penetrate and saturate Soviet air defenses without requiring the bomber itself to enter Soviet air space. The United States will also have more than 1,000 land-based ICBMs. True, these could be theoretically vulnerable in the mid-1980s when the Soviets have sufficient numbers of accurate ICBM warheads. The operational pitfalls in trying to destroy 1,000 launchers in a near-simultaneous surprise attack, however, make this theoretical vulnerability more fanciful than real. No matter what the Soviet Union does in the next ten to twenty years, the United States will still have overwhelmingly strong retaliatory forces; it cannot be disarmed.

It has been argued that Soviet defenses, and most recently civil or passive defense, could blunt a retaliatory attack and reduce damage to the civilian economy to acceptable levels. This thesis is unsupportable in light of the realities of nuclear warfare. Even under the most pessimistic assumptions on the extent of destruction from a Soviet first strike, the United States could retaliate with about 3,000 warheads from submarines at sea. Each of these would have yields at least three times that of the bomb that destroyed Hiroshima. Regardless of any civil defense preparations, these alone could knock out all Soviet aircraft on the ground, destroy more than two-thirds of Soviet industrial facilities, kill many tens of millions of people, and injure countless others. Contrary to a prevalent view, Soviet industry is more concentrated and thus more vulnerable than that of the United States; 200 of its largest cities contain 62 percent of its industry, and Soviet workers live close to their industrial buildings rather than in suburbs. Perhaps some machinery might remain undamaged, but the rehabilitation of the industry would be a colossal task, taking tens of years, because most of the wherewithal to put the economy back together again would have been lost.*

Although some people living in remote communities or managing to get into hardened blast and fallout shelters would survive, they would be restricted to scratching for bare survival in a ravaged environment made more unusable by widespread, long-lasting radioactive contamination. Evacuating urban populations into the countryside would not reduce their vulnerability to fallout, which can be alleviated only by staying in shelters for protracted periods. Furthermore the time required for evacuation would give the United States ample warning to allow it to increase significantly the scale of its retaliation. Radioactive contamination could be made more general by exploding many of the warheads close to the ground, a sound tactic in any case if one wished to destroy hard targets. There could be no possible Soviet gain from launching a first strike that would warrant risking the kind of destruction that could be visited on the Soviet Union in retaliation even using the most optimistic assumptions on the effectiveness of their civil defense.

Active defenses could potentially be more valuable in reducing damage. Soviet antiaircraft defenses are very extensive and

*A July 1978 CIA estimate concurred in by the entire U.S. intelligence community concluded that "Soviet measures to protect the economy could not prevent massive industrial damage." It also stated, "We do not believe that the Soviets' present civil defenses will embolden them deliberately to expose the USSR to a higher risk of nuclear attack."

could cut down significantly on the scale of our bomber retaliation, provided, of course, that they were fully effective at the time the bombers arrived. In any actual conflict situation, however, these defenses will have been severely weakened because of destruction by ballistic missile warheads that would have arrived much earlier. Furthermore, by the mid-1980s appreciable numbers of U.S. bombers will be equipped with twenty or more low-flying cruise missiles, which could saturate the air defenses and ensure widespread destruction of industrial and urban installations. They will also have sufficient accuracy to destroy hardened targets. In addition, Secretary Brown announced in September 1980 that the United States was developing radar countermeasures (the so-called stealth bomber) to cut down on the ability of Soviet defense systems to detect and track our planes.

Ballistic missile defenses have been so limited by the ABM Treaty of 1972 that currently they do not present a potential threat to the deterrent of either side. This treaty, which restricted both the size and technology of ballistic missile defenses, was an important step toward reducing fears that a real strategic imbalance might develop from ABMs. Even were the treaty to be abrogated, it would be many years before ballistic missile defenses could deal with the retaliation from the large numbers of MIRVed warheads that the United States would have available. A feasible ballistic missile area defense does not appear possible now, and any technological breakthroughs might threaten the deterrent only decades in the future. The widely publicized particle-beam weapons are nearer scientific fiction than reality, but laser-beam systems cannot be ruled out. They present, however, many very difficult and perhaps insoluble technological problems, may never be feasible as ABMs, and certainly are not a threat to the deterrent for many many years. Such systems cannot be deployed without amendment of the ABM Treaty, but research studies are now in progress.

These technologies should be watched carefully to see that they do not undermine the deterrent.

Similarly, the Soviet Union should have no rational fear of a U.S. disarming first strike. Their ICBMs are not vulnerable today, although they too could become theoretically threatened in the mid-1980s as the accuracy of U.S. missiles improves. As long as the U.S. ICBM force is limited to the 550 Minuteman III missiles, each with three MIRVs, and 450 Minuteman II missiles with single warheads, the total number of warheads (2,100) is marginal for it to be even a theoretical threat to the 1,400 Soviet ICBMs. If 200 of the larger MX ICBMs are deployed, each with ten MIRVs, then this theoretical threat would grow to the point that more than enough warheads will be available, but the same operational difficulties faced by the Soviets would preclude a realistic U.S. ability to carry out a first strike against Soviet ICBMs.

The Soviet submarine missile force is somewhat more vulnerable than that of the United States, since U.S. antisubmarine warfare in the Atlantic is far superior to that of the Soviet Union. Less than 15 percent of their submarines are at sea at any one time. Not only is our technology more advanced, but a large number of Soviet submarines, the Yankee class, are equipped with missiles with ranges of the order of 1,300 nautical miles, forcing them to operate in the Atlantic Ocean to be able to hit even the eastern United States. Here they are potentially vulnerable to U.S. antisubmarine warfare. This Soviet weakness will decrease as more and more of their submarines are equipped with missiles having ranges greater than 4,000 miles, thus being able to launch in the Arctic and the Far Pacific. Nevertheless even in such waters, the Soviets may be concerned about their submarine survivability since their ships are noisier and easier to detect and track than ours.

The Soviet long-range bomber force is far inferior to that of the United States since it is composed of only about 150 old aircraft, half of them propeller driven. The much-touted new Backfire bomber has limited range and is primarily designed for naval missions or operations in Europe or the Far East. It can reach the United States only on one-way flights and only by flying subsonically at high altitudes. It could be refueled in the air, but the Soviet ability to perform such operations is very limited. Even though the Soviet bomber force is old and slow, however, it could wreck major devastation in the United States because we have been phasing out our air defenses since 1973. The United States wisely decided that antiaircraft defenses were a waste of resources when we had no defense at all against ballistic missiles. Any air defense and probably the country would be destroyed by missile warheads long before their bombers arrived.

In sum, the strategic deterrent balance is very stable. Nothing on the horizon indicates that either nation could be disarmed by a first strike in the foreseeable future. Contrary to the many alarms that have been sounded, this balance is not delicate and cannot easily be upset. Even if one nation stopped all new weapons development and procurement and the other went ahead with all its programs without any restraint, it is hard to visualize an imbalance that could make a disarming first strike into a viable military or political action. The aggressor nation would be almost certain that it also would be destroyed as a modern society or at least reduced to third-class political status. Its political structure would collapse.

Limited First Strike

The security of the deterrent and the impossibility of a disarming first strike are not the only criteria by which the strategic balance is often evaluated. Few but the most extreme worst-case analysts would claim that the achievement of a capability to launch a disarming first strike and survive retaliation is a realistic goal.

Recognizing this impracticality, however, these worst-case analysts have proposed alternative war scenarios, which they claim the Soviets might use to exploit weaknesses in individual weapons systems. Thus attention has been focused on the potential vulnerability to a first strike of the land-based ICBM leg of our strategic deterrent triad.

Three factors are critical to the achievement of such an ICBM silo-busting capability. First, the attacking warhead must have the yield and accuracy combination of being able to destroy the hardened silo when it gets to the target (accuracy is much more important that yield). Second, the missile-launching system, from the command and control through the actual launching to the exploding of the warhead, must have a very high reliability of functioning as desired at a preset time. And third, a nation must have enough warheads with these characteristics to threaten the other nation's entire ICBM force.

Calculations that show a capacity of one country's ICBM force to destroy a large percentage of the opponent's ICBMs are a far cry from being able to rely on accomplishing this objective upon launching a first strike. Both former Secretaries of Defense Schlesinger and Brown have repeatedly stated that the extraordinarily complex operational difficulties in conducting such a strike would prevent any leader having confidence of being able to destroy in a first strike 1,000 or more opposing ICBMs. Unless a nation was certain of success, such an action would be most unlikely, since a failure would mean national suicide.

Although a theoretical countersilo capability cannot be practically relied on to destroy the ICBM portion of another nation's deterrent, it can be very destabilizing because of the appearance of a threat that it presents. The danger is multiplied many times when both nations have a potential countersilo capability. Then both perceive increased advantage from striking first and disadvantage from allowing the other to launch the first strike. These pressures for preemption should not be sufficient to induce any leader, Soviet or American, to launch an actual first strike against the other's ICBMs because even were it successful, the potential risks from retaliation by submarine missiles and bombers would be tremendous. Nevertheless it is a trend in the wrong direction, and a very destabilizing situation is created, since in time of acute crisis, such as a large-scale conflict in Europe, the pressure to launch could become almost intolerable.

Even more dangerous would be the incentive for one or both nations to adopt the tactic of being prepared to launch on evidence that a strike was underway. (This tactic is known as launch-on-warning or launch-from-under-attack.) This approach is already being given new credence in the United States as fears of the Soviet threat mount. Such an operation is technically feasible because modern missile systems can be designed for launch on a few minutes' notice, while detection systems, such as radars and infrared sensor equipped satellites, can give a minimum of fifteen to twenty minutes warning that an attack has been launched. Thus a very cheap and completely reliable way to prevent a nation's missiles from being destroyed by a countersilo capability would be to adopt such a launch-on-warning posture.

This tactic is extremely dangerous. It would greatly increase the risks that a nuclear conflict could break out by accident. Radars and computers can fail and create false alarms; in this case a false alarm could mean Armageddon. Indeed three computer failures led to false alerts of a Soviet missile attack in November 1979 and early 1980. In those cases the dangers of an actual nuclear response were low because we had not adopted a launch-on-warning posture. Had we done so, the risks of accidental war would have been very much higher. Mutual countersilo capabilities, even when perfectly balanced, can be extremely precarious because they might induce either or both nations to adopt this risky tactic.

The whole concept of a balance in countersilo capability is irrational and should not be used as a criterion for the adequacy of strategic forces. A countersilo capability is a goal that neither nation should seek regardless of the action of the other since it can only increase the risks of a nuclear disaster. Thus even if the Soviet Union does move in this direction, our answer should not be to match it. This would be a mindless response. It is regrettable that the new defense secretary, Caspar Weinberger, defended the MX as needed to match the large Soviet countersilo missiles. This position can only increase the likelihood that the Soviets would actually fire their countersilo weapons, with catastrophic consequences for all. This is an example of how a balance in strategic force capabilities can create dangerous incentives for initiating a nuclear conflict.

It is time to cease having our actions dictated by unreasoned fears and to recognize that the launching of a countersilo first strike against the ICBM portion of the deterrent is not an option for any nation's leader since a disarming first strike is not possible. The theory that the United States would not respond under such circumstances with submarine missiles or bombs for fear of subsequent Soviet strikes is not one that any Soviet leader could rely on. Were his judgment to prove wrong, it would mean the end of his country as a society. There can be no possible Soviet military gain from destroying our ICBMs that could warrant taking the risks entailed in such aggression. The chance that a U.S. president would forgo retaliatory action after thousands of warheads had exploded on U.S. soil and killed tens of millions of people is not high, even though such a response might result in further devastation to the United States.

Furthermore, whether we had or had not a countersilo capability would be immaterial since this would be ineffective in a second strike. If anything, the availability of such a capability might mislead the Soviets into believing we might retaliate in this ineffective fashion. For then Secretary of Defense Schlesinger to imply, as he did in testimony before the Senate Foreign Relations Committee in 1974, that we might not respond after a Soviet attack on our ICBMs was extraordinarily damaging to our deterrent posture, the essential element of our strategic security today. General George Seignious, former director of the Joint Staff of the Joint Chiefs of Staff, testified in 1979, "I find such a surrender scenario irresponsible—for it sends just the wrong message to the Soviets. We have not built and maintained our strategic forces—at the cost of billions—in order to weaken their deterrent impact by telling the Russians and the world that we would back down—when, in fact, we would not."

Perceptions of the Strategic Balance

This leads directly to the question of how the strategic balance is perceived at home and abroad. Because a disarming first-strike capability is not an achievable goal for either the United States or the Soviet Union in the foreseeable future and because even a successful countersilo attack against the ICBM portion of the deterrent could risk the annihilation of the aggressor's society and therefore is not a viable action, one is left only with the consideration of whether any perceived strategic imbalances can be used for political gain. Perceptions are difficult to evaluate because they depend on the information made available to the world and the form in which it is presented. But facts can also easily be misrepresented, and often fears of a perception can become self-fulfilling prophecies. This situation currently is driving the arms race and creating roadblocks to arms control agreements.

In order to determine whether such perception could now or in the future lead to a situation that would work to the political disadvantage of the United States, it is necessary to see what kind of imbalances now exist or might do so in the future. Because the U.S.S.R. and U.S. weapons programs have been molded by different military, geographic, and political factors, there are many asymmetries in their respective force structures. Historically the Soviet Union has elected to procure heavier, less sophisticated missiles, while the United States has gone the route of smaller but higher-technology weapons. This shift, which began in the 1950s when the Soviets moved from bombers to ballistic missiles before the United States, occurred before hydrogen warheads were light enough to allow anything but very large missiles to carry them to intercontinental ranges. By the time the U.S. program reached the hardware stage, lighter payloads and thus smaller missiles were made possible by more advanced weapons technology.

Another example of where the two nations followed different courses was MIRVs. In the mid-1960s the United States elected to halt the deployment of additional missile launchers and instead to increase its force levels by placing many MIRVs on a single missile—three on its Minuteman III ICBMs and up to fourteen on its Poseidon submarine missiles. This strategy was defended by then Deputy Secretary of Defense Paul Nitze before Congress in 1967 as the most effective way of expanding our strategic capability. The Soviet Union, lagging the United States in MIRV technology by about five years, continued to deploy both ICBMs and submarine missiles with single but higher-yield warheads. Only in the last six years have they begun to deploy missiles with MIRVs, and even then not on all of their new ones.

Thus the United States has always had a lead in warheads even though we had about two-thirds as many missiles as the Soviet Union. This lead in warheads is now vanishing as the Soviets put MIRVs on their large missiles. The United States still has many more bombers, each carrying a larger number of weapons, so the United States now has a total of 9,000 strategic warheads and bombs to 7,000 for the Soviet Union. Since it would be hard to find 9,000 legitimate targets for nuclear weapons as long as there are no ABMs to overwhelm or Soviet MXs in multiple shelters to destroy, this lead cannot easily be turned to any political advantage; indeed the United States has not sought to do so.

U.S. warheads have been more accurate than those of the Soviet Union, but this is important only if a countersilo capability is being sought for a first strike. The Soviet Union, on the other hand, has higher-yield warheads, but this is not particularly significant since most targets do not require the overkill of a multimegaton explosion. Accuracy is much more important than yield for attacking hard targets. Even this characteristic has vanishing significance since once pinpoint accuracy—the ability to hit within 100 yards or less—is attained, there is little gain from further improvement in accuracy because even relatively small nuclear warheads can destroy very hard targets at that distance. Then reliability takes over as the key parameter. Thus both nations have leads in certain missile characteristics, which can be selectively cited to prove the United States is ahead or behind, but it is hard to see how any of these asymmetries can be used to obtain political gains unless we wish to advertise falsely our weaknesses.

In the strategic bomber area there exists a major asymmetry resulting from the Soviet decision in the mid-1950s to shift from bombers to missiles as the mainstay of its strategic forces. As a consequence they have not built any new true intercontinental bombers and are left with a force of 156 old Bison jet and Bear turboprop aircraft. Secretary Brown admitted in January 1981 that they "had been expecting the Soviets to develop a new long-range bomber for several years," but it has not yet appeared. The much-advertised Soviet Backfire intermediate-range bomber is not a good strategic system, and the Soviets would be much stronger militarily if they kept this type of bomber for their designed naval and peripheral European and China missions.

The United States, on the other hand, chose to stress its intercontinental bomber force and still has about 350 B-52s fully operational and many others in standby status in the aftermath of the Vietnam war. The United States is modernizing this bomber fleet with many long-range cruise missiles for launching weapons outside the Soviet defense perimeter to make it easier to overwhelm the defenses. This is a more effective and cheaper means of maintaining the viability of the aircraft element of the strategic triad than a new penetrating bomber, such as the B-1 or the other planes being considered by the Reagan administration.

This U.S. lead in strategic bombers has been largely ignored and not perceived as a major imbalance in the U.S. favor. Nevertheless it could have been used to proclaim U.S. superiority and could again in the future when the United States deploys thousands of accurate cruise missiles for launching from its bombers. However, the United States is not prone to advertise its strengths. On the contrary we seem to prefer to emphasize our weaknesses, perhaps to enlarge the budgets for more weapons. This is an unfortunate tactic because in many areas the United States has and will continue to have a lead that can be cited to give the perception of U.S. advantage if perceptions worry us.

In one category of weapons the United States is commonly perceived to be at a disadvantage: land-based ICBMs. This has dominated most attention in the political-military debate and in the press. The United States has put about a third of its total strategic strength and the Soviet Union about three-quarters into land-based ICBMs. As a result the Soviets now have more ICBMs than we have (1,400 compared to 1,052). They also have larger ones so that with or without the MX, they eventually can have many more ICBM warheads than we (table 7.1). Their SS-18s have been tested with ten MIRVs each and their SS-19s with six, but in the absence of SALT limits, further development could double or triple the number of MIRVs each can carry. (For comparison, our Minuteman III carries three warheads but has been tested with seven.) Now the Soviets are said to be acquiring an accuracy for these missiles comparable to that which the United States has had for several years, and this will soon give them a theoretical threat to our Minuteman silos. The reliability of their ICBMs is reported to be inferior to that of the United States; if so, this should reduce their confidence in the success of any countersilo strike. Nevertheless it is the perception of this countersilo capacity that is used to justify the new multiple protective shelters program for the deployment of the MX.

This Soviet ICBM program is being widely used to foster a perception of U.S. inferiority; all of the other factors in the strategic equation are being ignored. Even were our ICBMs to become vulnerable, they are less than a third of our deterrent. Even without the MX, the United States is also procuring 1,000 to 1,500 advanced Minuteman warheads that can threaten a significant fraction of the 1,400 Soviet ICBMs, the major element in their deterrent. What is really a U.S. strength—the approximately equal division of strategic forces among ICBMs, submarine missiles, and bombers—is being labelled a U.S. weakness by isolating the ICBMs and ignoring the others. What was really a serious Soviet mistake—the concentration of its forces in the now increasingly vulnerable ICBMs—is being passed over and in fact touted as a Soviet advantage. This is an example of how perceptions can be dangerous and too often can be manipulated to promote narrow interests (figure 9).

A broad analysis of the strategic forces shows that the United States has and will continue to have a balanced, survivable, and effective strategic force, which in its overall deterrent characteristics is far superior to that of the Soviet Union. With its accurate ICBMs and advanced antisubmarine warfare, the United States can pose a threat to the Soviet deterrent as a whole, far greater than that of the Soviet Union to the United States. Thus there is no sound basis for a perception of U.S. strategic inferiority.

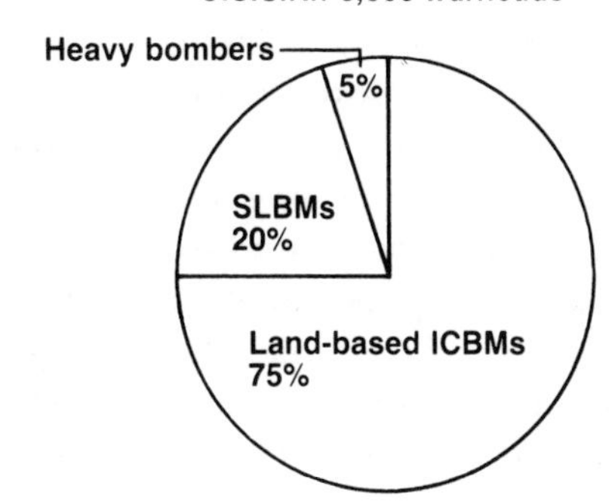

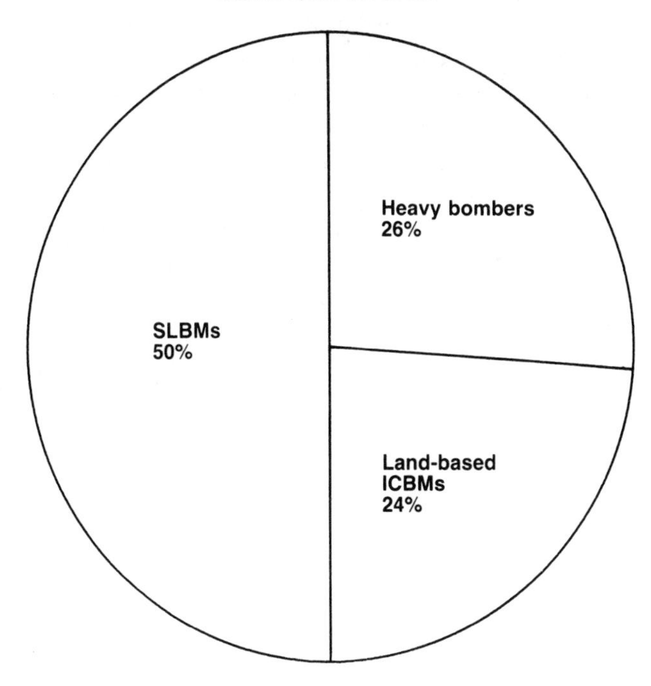

9. Composition of U.S. and U.S.S.R. strategic deterrent forces in 1980 (DoD). The United States with its advanced Minuteman III ICBMs can threaten to destroy about 50% of the Soviet strategic force (3,000 warheads), while the Soviet Union can threaten to destroy less than one quarter of the U.S. force (2,150 warheads) even if its ICBMs could destroy all U.S. land-based missiles.

This analysis demonstrates how easy it is to create a misperception of relative strategic strength by isolating specific comparisons from the overall balance. More importantly no comparisons have any real meaning as long as a disarming first strike is not even theoretically possible and as long as a counterforce attack against the ICBM component of the deterrent runs a suicidal risk that far outweighs any potential gain. The force levels of both the United States and the Soviet Union are so high in all categories of weapons that any inequalities have little significance. Seeking a balance in countersilo strengths is the ultimate madness since it increases incentives to start a nuclear conflict. Denigrating our military strengths and spreading unbalanced perceptions of our weaknesses and our intentions is contrary to our national security because it undermines our deterrent. The debate over who is ahead has lost all perspective as the players have forgotten the catastrophic damage that even a few nuclear explosions can produce. The debate must be reoriented toward more meaningful issues, and our concerns over the strategic balance redirected to a concern over the consequences to all humanity if nuclear weapons are ever used again.

9 SALT and the MX

While the United States and the Soviet Union were building up their strategic arsenals in the 1960s and 1970s, leaders in both countries were recognizing that arms control was a complementary and potentially useful alternative approach toward security. Many of them understood that without any restraints, the deterrent balance could be upset by new types of weapons and the cost of maintaining a stable balance would skyrocket.

It was one thing to recognize the importance of arms control, however, and another to achieve agreement on practical measures. Neither country wished to freeze itself into a position of apparent inferiority, and yet both wished to press their advantage in the areas in which they had an apparent lead. Ironically if one were ahead, limitations were often unattractive and if one were behind, they were unpalatable; only at the rare times when the forces of the two sides were viewed essentially equivalent were agreements considered practical.

As a result of the rapid buildup in strategic delivery vehicles in the early 1960s, the United States had a commanding lead over the Soviet Union. Russian leaders instituted vigorous programs in the middle and late 1960s to redress this imbalance. Although they had accepted in principle the idea of limits on offensive and defensive strategic weapons, it was not until they could see that their weapons programs were well on their way to completion that they were willing to enter into serious negotiations to limit arms buildups. Initially they were particularly reluctant to address ABMs. The United States, on the other hand, was looking for an opportunity to consolidate its position in offensive missile launchers, but not on MIRVs, which were in the research phase. It simultaneously gave high priority to limits on Soviet ABMs, viewed as a potential threat to our deterrent. By 1968 the Soviets could see clearly the time when they would have as many or more strategic missiles as the United States and at the same time could realize that their ABMs would not provide an effective defense. Therefore they were willing to enter into specific strategic arms limitation negotiations on both offensive and defensive weapons. The actual start of the Strategic Arms Limitation Talks (SALT) did not occur until November 1969; the talks scheduled to begin in the fall of 1968 were deferred because of the Russian movement of troops into Czechoslovakia.

By the time SALT got underway, the United States had changed its attitude toward ABMs because it had designed the Safeguard system to defend its Minuteman silos. Although Safeguard would not have provided any real protection since it could be easily overwhelmed by the existing Soviet forces, the Nixon administration was no longer willing to put ABM controls ahead of those on offensive weapons. By 1970 the number of Soviet ICBM launchers had exceeded that in the United States, and we were interested in ensuring that this Soviet buildup would not go on indefinitely.

Although Secretary of Defense McNamara had called a halt to building additional strategic missile launchers in 1965, he had turned to MIRVs as the more cost-effective way to enlarge strategic force levels. The first test of a U.S. MIRV occurred in 1968, and deployment was scheduled to begin in 1970. Although at that time a consensus in the United States was very much in favor of controlling the number of strategic missiles and their launchers, there was a vigorous debate as to whether MIRVs should be similarly controlled. The United States was about five years ahead of the Soviet Union, and some were reluctant to put restrictions on a program that took advantage of U.S. technological superiority. On the other hand, many others foresaw that in time Soviet MIRVs, if left uncontrolled, could threaten our land-based missiles and result in a serious loss to our security. President Nixon's national security adviser, Henry Kissinger, had never favored MIRV limitations in SALT I; however, the prestigious Presidential General Advisory Committee on Arms Control, headed by John J. McCloy, urged the inclusion of MIRV restrictions in SALT I. The Defense Department, sparked by recommendations from Paul Nitze, its representative on the SALT negotiating team, and by the views of the Joint Chiefs of Staff, strongly opposed any inclusion of MIRVs in SALT I. Opponents of MIRV limitations argued that we were ahead of the Russians in MIRV technology, that technology was our specialty, and that therefore we should not put restrictions on developments in which we were better than the Russians.

In 1970 the United States did make a proposal to ban MIRV testing and deployment but insisted, contrary to the McCloy committee recommendation, on on-site inspection for verification purposes. Any practical inspection scheme would not significantly help the verification of MIRV deployments, however. Unless inspectors were permanently stationed at every missile launcher, something unacceptable even to the United States, it would be too easy to substitute a single warhead for the MIRVs in the final stage of the missile before the inspector arrived at the site. Since it was also known the Russians would not accept on-site inspections, this U.S. offer was viewed as a patently nonnegotiable ploy that gave the appearance of being interested in arms control. In SALT II, satisfactory arrangements were worked out for verifying MIRV limits without on-site inspections; satellites and other technical intelligence means could observe missile tests and deployment areas to determine which missiles had a MIRV capability.

The Soviets responded with a proposal to ban production and deployment but not testing, from which they were still two years away. This offer too was nonnegotiable since controls on tests are essential for limiting the development and later deployment of new type of missiles. Only through observations of tests could we verify the status of the Soviet MIRV program and determine whether MIRV missiles were being deployed.

The United States never sought to negotiate seriously the differences in position of the two countries because the government had already decided not to seek a MIRV ban in SALT I.

This was a critical mistake for which we are paying dearly today; now it is the Soviet MIRVed missiles that are threatening our ICBMs and are being used to justify the MX deployment in its multiple protective shelters. We would not be worrying now about the vulnerability of our Minuteman ICBMs and talking about spending tens of billions of dollars to overcome it if we had succeeded in limiting MIRVs in SALT I. By the time of SALT II it was too late. This is a classic example of how the decision to seek security through an arms race with the Soviet Union instead of through arms control has had a profoundly negative effect on our security. And experience has shown how hard it is to turn the clock back.

SALT I did succeed in achieving the ABM Treaty, which limited the ABMs on both sides to two sites, one protecting the nation's capital, similar to the Soviet Moscow system, and one protecting an ICBM site, similar to our Safeguard. Neither of these two ABMs would have had any significant military value, but the agreement to allow them satisfied the political requirements of the two countries. In a Protocol to the Treaty, signed two years later in the dying days of the Nixon administration, each side agreed to reduce ABM deployments to only a single site, thereby giving tacit recognition of the ineffectiveness of these weapons. The United States never made its one site fully operational, and the Soviet Union, some ten years later, has never completed even its small deployment around Moscow. The treaty also choked off many possible future destabilizing ballistic missile defense technologies.

The ABM Treaty was a major security gain since it assured for the foreseeable future that any warhead launched in a retaliatory attack could travel unhampered on its way to its intended target. Thus mutual deterrence was guaranteed. In its security significance the ABM Treaty far outweighs any other arms control agreement achieved since the advent of nuclear weapons in 1945. It has done more to enhance our security and reduce the risk of war than all the thousands of new missiles and warheads we have procured in the last ten years at a cost of billions of dollars.

Logically the ABM Treaty should have halted or at least slowed the offensive weapons race as well, since it is hard to make any rational argument that more warheads are needed on either side when there are no defenses to be penetrated. But rationality did not dictate either nation's defense programs; both went ahead with offensive weapons development and procurement at an accelerated rate. We doubled the number of our missile warheads after 1972. The new weapons, particularly MIRVs, were very destabilizing, increased the likelihood of a nuclear conflict's breaking out, and gave a new impetus to the arms race. It is these developments in offensive weapons since 1972 that led to the MX program decision in 1979.

Although no controls were established on MIRVs or on any other means of modernizing offensive delivery vehicles, SALT I did succeed in negotiating an interim agreement lasting five years and putting limits on certain offensive weapons. The numbers of missiles on both sides were essentially frozen at their 1972 levels. The Soviet Union, however, was allowed to replace its obsolescent ICBMs with new submarine-launched missiles that permitted a substantial enlargement of that part of their deterrent.

SALT II negotiations were begun shortly after the SALT I Agreements were ratified with a view to achieving at an early date a permanent treaty limiting offensive weapons. Regrettably such a treaty was more difficult to conclude than had been originally hoped. In the meantime the political climate and public support for arms control underwent a major change. Although there was no difficulty in getting the SALT I agreements ratified, the administration lost control of the debate to Senator Henry Jackson, a long-time opponent of arms control. The administration spent its efforts defending why the agreements were not bad rather than emphasizing their positive benefits. Senator Jackson succeeded in getting passed a Senate resolution that mandated that any future arms control agreement did not leave the United States in a position of numerical inferiority. Although no policy maker, arms control negotiator, or senator who valued reelection would support an agreement that purportedly left the United States inferior, this resolution nevertheless set the tone for an insistence on a strict equivalence of forces on both sides no matter how irrelevant such a balance was to our security. Since there were many asymmetries in the postures of the two countries, as well as in the geography, it proved hard to achieve limitations that satisfied everyone's standard for equivalence.

The departure of President Nixon and the assumption of the reins of the government by President Ford, who had not had any previous involvement in arms control, further delayed progress. At Vladivostok in the fall of 1974, Ford and Brezhnev reached an accord on the framework for a treaty that would have given both sides roughly equivalent forces. The translation of this broad framework into specific treaty language, with all the details pinned down proved extremely difficult, however. Governor Ronald Reagan made his opposition to SALT a strong point in his 1976 primary campaign, and as a result President Ford backed off from any agreement, even though Secretary of State Kissinger had worked out the mechanics for the controversial but critical problem of limiting MIRVs.

The advent of the Carter administration still further delayed progress. It attempted to scrap the Vladivostok Accords and on March 30, 1977, quickly moved to a much more far-reaching proposal without first laying the proper groundwork with the Soviet Union. Although many of the new measures could have been extremely valuable, the March proposal contained many one-sided items obviously unacceptable to the Russians. National security adviser Zbigniew Brzezinski unsuccessfully attempted to rationalize the provisions favoring the United States by claiming that the proposal would also stop the MX missile, which he described then as a first-strike threat to Soviet ICBMs. The attempt to achieve quickly very extensive limitations backfired, and the SALT process was seriously slowed. The eventual ratification of SALT II was made more difficult because it looked as if we had retreated from our initial position under Soviet pressure.

When the negotiations finally got back on the track in the late summer of 1977, MIRVs were the key weapons that needed to be dealt with. The danger that they presented to our ICBM forces was at last recognized. Arrangements were worked out on limitations that did not require on-site inspections for verification of compliance.

Although this thorny issue had been essentially solved, a new difficulty arose that interfered with the achievement of a significant arms limitation agreement in SALT II. As the MIRV threat to our Minuteman ICBMs became more and more apparent, weapons researchers tried to solve the problem through a replacement missile system, the MX. This new weapon began to acquire a life of its own even though no satisfactory solution could be found for it to ensure ICBM survivability in light of the anticipated improvements in Soviet MIRV capabilities. Those who believed in the weapons solution to security as opposed to arms control insisted that any SALT Treaty must allow the MX program to proceed. Ironically the same individuals and groups that had insisted that no controls be placed on MIRVs in SALT I were now vigorously warning of the dangers to our Minuteman ICBMs from Soviet MIRVs. They were also criticizing the draft SALT II treaty for not eliminating this MIRV threat they had allowed to spawn.

Of course the Soviet Union was also going ahead with its new weapons development programs, and the SALT negotiators were faced with the problem of trying to deal with rapidly changing situations on both sides. A complete version of the treaty was finally achieved in the spring of 1979. On the whole it was a very useful contribution to our own and world security; it made the strategic balance firmer, started the process of reducing force levels, and put some restrictions on new technology. But it was still a long way from satisfying the desires of those who were looking for a treaty to bring the arms race to a halt.

The most significant weaknesses in the treaty were in the area of restrictions on new types of weapons. Although at one point the Soviet Union was willing to ban the testing and deployment of all new types of ICBMs, the United States, in order not to foreclose the development of the MX, insisted that each nation be allowed one new model. In SALT I the United States, under Nixon and Kissinger, had sought to ban all mobile ICBMs, but the Soviets had adamantly opposed any such restrictions. Now as SALT II reached its final stages, the positions of the two countries were reversed. The Soviets were willing to give up all mobile ICBMs, but the United States insisted on keeping this option open for the MX. This insistence on allowing each side to proceed with a single new type of ICBM, and a mobile one at that, resulted in the biggest loophole in the SALT II Treaty and one that the U.S. may come to regret, just as it now regrets not having limited MIRVs in SALT I.

By the time SALT II negotiations were well underway, it was too late to stop MIRV programs at sufficiently low levels that neither side would have enough warheads to threaten the ICBM deterrent of the other. The United States had 550 MIRVed ICBMs and 496 MIRVed SLBMs; in 1977 the Soviet Union was still considerably behind, with only about 300 MIRVed ICBMs and virtually no such SLBMs. The Soviet Union, however, was not about to agree to a treaty that left it in an inferior position concerning such an important type of weapon, and the United States was not about to scrap its newly acquired MIRVed missiles. Therefore although the SALT II Treaty did put an overall ceiling of 1,200 on MIRVed missiles of all types and a subceiling of 820 on MIRVed ICBMs, these levels were too high to solve the MIRV threat problem. That opportunity had been irretrievably lost in SALT I.

The treaty did, however, break some important new ground in the area of controls on total numbers of warheads by placing an upper limit on the number of warheads each type of missile could carry. Because the Soviets had tested their very large SS-18 with ten warheads, this was the maximum that they would be allowed to deploy. The number of warheads released in any one test could easily be verified by observation of Soviet test firings, particularly as the warheads reentered the atmosphere over the Kamchatka Peninsula or the Pacific Ocean. Thus there need be no fear that the Soviets were cheating on this provision. Similar limitations were placed on the warheads in each of the other types of Soviet missiles and also on those of the United States. Although the United States had never tested an ICBM with ten warheads, it would be allowed to have ten warheads on the single new ICBM it was permitted to develop under the treaty. Since this new ICBM was to be the MX, this accounts for why the MX is designed to carry ten warheads.

By assigning to the entire Soviet ICBM force the maximum number of warheads each type of missile was allowed to carry, the total number of warheads that the Soviets could have under the treaty could be calculated, as shown in table 9.1. This total number of warheads led to the original decision that the MX basing scheme should include 4,600 protective shelters. The administration believed that with a maximum of about 6,500 warheads allowed under SALT II, the Soviets could never rely on destroying 4,600 hardened shelters. Thus the SALT agreement was an essential element in ensuring the effectiveness of the proposed MX deployment. This is a good example of how closely meshed the MX program was with the SALT II Treaty. Yet many MX enthusiasts have been the strongest opponents of SALT.

Table 9.1
Soviet ICBM warheads allowed under SALT II

Type of missile	No. of missiles	No. of warheads
MIRVed ICBMs		
SS-18s with 10 warheads	308	3,080
SS-19s or SS-17s with 6 or 4 warheads, respectively	512	3,072 (max.)
ICBMs		
Missiles with 1 warhead	380	380
Total	1,200	6,532 (max.)

The expiration date for the SALT II Treaty, had it come into force, would have been the end of 1985. Any extension would have had to be renegotiated in light of the existing security situation at that time. The existence of a firm MX program calling for 4,600 shelters would make it very difficult for the Soviets to agree to its continuation into the indefinite future. They would almost certainly oppose any treaty that banned them from increasing the number of their warheads to deal with the 4,600 shelters. Furthermore the SALT II Treaty has not yet been ratified, and there is not much optimism that it or a revised version will be ratified in the near future. To date both nations are abiding by the basic provisions of the SALT II Treaty, and it is in our interest that we continue to do so. At any time, however, the Soviet Union could test its SS-18 ICBM with more than ten warheads or its SS-19 with more than six. Then the limitations on warheads, which are now in the agreement, could fall apart—perhaps like Humpty-Dumpty, never to be put back together again. These considerations are vital as decisions are made to spend tens or even a hundred billion dollars on the MX missile system and we defer action on further strategic arms limitations.

10 The 1979 Decisions

Although the SALT II Treaty provided the final impetus to the MX decisions in 1979, the genesis of the MX missile program itself goes back at least to 1967 when the Defense Department sponsored an investigation of a broad range of alternative weapons to replace the just completed Minuteman and Polaris-Poseidon missile systems when these eventually become obsolete. This study, which was carried out by the department's in-house Institute for Defense Analyses, was known as Strat-X and became the bible for future programs for the next ten years. It analyzed a wide variety of missile concepts, ranging from mini- to very large ICBMs, and of deployment modes, including mobile systems that would travel along the ground, underground, in the air, and in and under the sea. In the early 1960s, mobile basing had been considered for the original Minuteman ICBM (moving the missile from place to place on roads or rail) but had been discarded in favor of placing the missiles in fixed hardened silos. The Trident submarine was the first of the new Strat-X weapons to be put into actual production, but many of the other concepts were later studied in detail when the MX missile replacement for the Minuteman underwent serious consideration. The idea of moving Minuteman missiles around the roads in their present basing areas has recently been resurrected but still found wanting because the missiles are too vulnerable when exposed in the open.

Although they are necessarily interrelated, the MX decisions should be separated into two parts: the design and characteristics of the missile itself and the mode in which the missile would be based. That these are two separable considerations was demonstrated by President Carter when he made his decisions on the MX in 1979. He first selected the characteristics of the missile that was to be built and then six months later chose the way that the missile would be deployed.

The MX Missile Decision

As the decision-making process for the MX missile moved to a close in 1978 and early 1979, the Carter administration faced two conflicting basic design choices for the missile itself. Should the missile be small enough to make it compatible for launching from the newly constructed Trident submarines as well as from land, or should the missile be the largest ICBM that would be allowed under the SALT II Treaty, then in final negotiation? A common missile that could have been used by both the navy and the air force would have had many advantages because if a decision were later made to move the force from land to sea, this could be accomplished without building an entirely new missile. It could still have carried ten warheads, and it would have been easier and cheaper to get a survivable mobile land basing system. If the missile were ever to be air mobile (transportable from one place to another by aircraft), then the smaller size would be needed. The common missile also would have reduced the requirement for an additional new submarine-launched ballistic missile, the Trident II, which was planned to provide the navy at a later date with a long-range, hard target weapon. Neither the air force nor navy, however, was particularly entranced with the idea of a common missile, which divided management control and might end up by each one sacrificing some of their particular desires. The experience in the McNamara regime with the TFX aircraft, which was supposed to give the air force and navy a common fighter plane and ended by satisfying neither, gave substance to this skepticism.

The other option—to build the largest missile possible within the SALT II limitations—had many supporters in 1979, for both political and military reasons. The inflated concerns over the Soviet large missiles, which had been fostered during the SALT II debates, created many political pressures in this direction. Also the larger the missile, the easier it would be to satisfy desires for a silo-busting capability.

From the very earliest days of its ballistic missile program, the United States had always developed and procured smaller missiles than had the Soviet Union. This option was made possible by the more advanced technology of both delivery systems and nuclear warheads in the United States. Our first-generation Atlas and Titan missiles were considerably smaller than the Soviet SS-6. Only relatively small numbers of Atlas and Titan missiles were ever deployed because even these were large. The second-generation Minuteman was considerably smaller than either and safer to maintain and launch on short notice.* The relatively small Minuteman III with three MIRVs later became the backbone of the U.S. land-based ICBM force in the 1970s and will continue to be so in the 1980s.

The Soviet Union, on the other hand, had to opt for much larger ICBMs. Even their second- and third-generation missiles were considerably bigger than those of the United States. The SS-9, of which they had deployed 308 by 1972 when the program was frozen by SALT I, was capable of launching a single warhead of about 20 mt. Even its lighter counterpart, the SS-11 ICBM, was larger than the Minuteman. Although the accuracies obtainable with these systems were distinctly inferior to those of U.S. missiles, it was the very large payload, or throw weight, of the Soviet missiles that was cited as evidence of Soviet superiority in ICBMs during the SALT I and SALT II debates. The throw-weight gap became a major political issue because the United States under both the SALT I Interim Agreement and

*The Titan used storable (not cryogenic, requiring refrigeration) liquid fuels, and these caused the accidental explosion in September 1980. Two men were also killed in an earlier Titan accident, and fuel leaks have occurred on numerous other occasions. Even today virtually all Soviet missiles, including its submarine missiles, are liquid fueled.

the SALT II Treaty was precluded from procuring heavy missiles similar in size to the SS-9. This controversy was heightened by the Soviet testing and eventual deployment of even larger payload missiles, the SS-18 replacement for the SS-9 and the SS-19 replacement for the SS-11.

This heavier payload gave the Soviets the potential to have a much greater number of ICBM warheads with higher yields than the United States. Although initially U.S. guidance systems were far superior to the Soviet Union's, it was foreseen that in time the Soviets would also achieve this technology and thus eventually could have a greater hard-target capability than the United States. Hawks in Congress and outside the government put tremendous political pressure on the executive branch to move to larger missile systems. The Defense Department, however, did not see the need for and never established a military requirement for a missile equivalent in throw weight to the Soviet SS-18. The U.S. choice to procure the smaller Minuteman ICBM instead of the larger Atlas and Titans was a deliberate decision by military planners.

The comparative size of U.S. and Soviet missiles nevertheless has continued to be an important political issue. As the SALT ratification debate approached, the question of big missiles became more and more of an emotional argument. Under the SALT II Treaty both the United States and the Soviet Union could develop one new ICBM of the same size as the current Soviet SS-19 but no larger. Because of the close connection between SALT II and the MX, the demands on the Carter administration to select a missile with this largest permissible size were very strong. A somewhat smaller design with a diameter of eighty-three inches instead of ninety-two could have made the new missile launchable from existing submarine launch tubes, and William Perry, under secretary of defense for research and engineering, testified this could be built with "no significant degradation of performance to the MX missile." It could still carry ten of the first-generation 335 kt MX warheads. These considerations, however, were smothered by the cries for big missiles.

In the spring of 1979, political demands to take some action on new weapons mounted steadily. Although several hundred million dollars were being spent each year on the MX research and development, the basic decisions to proceed were still in the indefinite future. The Carter administration had been under serious and growing criticism for its cancellation of the B-1 bomber and for the delay in making a commitment to the full-scale development and deployment of a new MX missile system to replace Minuteman. Claims were made that our security was being jeopardized and that the Soviets soon would be in such a superior position that the United States would be considered a second-class power. Critics ignored the fact that all proposed MX deployment schemes would not have worked to reduce ICBM vulnerability.

At the same time the SALT II Treaty negotiations were reaching their final stages, with only a few details left to be worked out. As the treaty was delayed longer and longer, however, the oppostion to it was growing by leaps and bounds. Without any final version of the treaty being public, the administration was having difficulty in defending it against attacks by those who argued against selectively leaked individual provisions rather than considering the treaty as a whole. Soviet actions around the world increasingly were being cited as reasons why it was not timely to conclude a SALT II treaty. Senators, who were skeptical about SALT and who were considered important for SALT ratification, made it clear that unless the MX program was irrevocably approved, the administration could not count on their backing for the treaty. They did not go so far as to say that they would vote for SALT if the MX was authorized, but they did say that without such approval, there was little chance that they would support it.

Thus with the SALT II Treaty all but wrapped up, the Carter

administration finally decided to bite the MX bullet. On June 8, 1979, the White House announced that the 192,000-pound, 92-inch-diameter design, the largest possible version of the MX missile, had been selected for full-scale development. The idea of a common air force–navy missile was discarded. This announcement was made only eleven days before the SALT II Treaty was signed in Vienna. Despite White House protestations to the contrary, there is little doubt that the decision was dictated by the desire to facilitate the ratification process. The large MX missile was one of many weapons used as bargaining chips to attempt to buy Senate votes for SALT—not to obtain Soviet concessions.

In June 1979, Carter decided that the higher-yield Minuteman warhead with its advanced guidance system would be incorporated in the first version of the MX and that then this would be still further improved in subsequent modifications. The entire system would be composed of 200 missiles, each with ten warheads, so it will have a total of 2,000 accurate, high-yield warheads. The Soviets have no choice but to consider this a direct threat to their 1,400 Soviet ICBMs.

Ironically the politics of the SALT II Treaty, which was designed to limit strategic weapons and which had been under negotiation for seven years under three presidents, led President Carter to make the decision to proceed with full-scale development of the new MX missile, the most dangerous, destabilizing weapon and greatest impediment to meaningful arms limitations in a decade. It was not the first occasion that arms control had given a boost to a weapons program, but the MX was certainly the most obvious and dramatic example. But unlike earlier cases, arms limitations are necessary to make the MX deployment effective so the final ironic twist may be that arms control failures may lead to a reversal of President Carter's MX decisions.

The Basing Decision

The primary requirement for a basing plan for the MX or any future land-based missile system is survivability. ICBMs should not be allowed to become attractive targets for any first strike. A significant fraction of the force should survive for retaliation if that leg of the deterrent triad is to remain viable. Although it was recognized that no Soviet leader could be assured of a successful surprise attack against the entire force, vulnerable ICBMs are destabilizing and could increase the risk of a nuclear war.

Many of the proposed plans, including multiple protective shelters, depend for their effectiveness on a reliable knowledge of the number of Soviet missiles and warheads. If the Soviets have significantly more warheads than we have shelters, survival of the missiles cannot be assured. Thus a second requirement for any basing was that it must provide a credible way for the other side to verify how many missiles of a given type are actually deployed in launchers. Even in the absence of an arms control agreement this is desirable, but in a treaty situation this is essential. Verifiability was easy for fixed land-based ICBMs because the launchers could easily be seen and counted by satellite cameras. This becomes more difficult for mobile ICBMs. In the basing design for the MX the United States went to extreme lengths to come up with some plausible scheme that satisfied this requirement.

Until 1972 there was little support in the United States for basing an ICBM in a mobile mode. During the SALT I negotiations, we attempted to get Soviet agreement not to replace existing land-based missiles with mobile versions. We feared that our ability to verify the size of the Soviet force would be degraded because of the difficulty in counting accurately the number of mobile missiles deployed. With fixed basing it was not difficult to make an accurate tally on the number of Soviet launchers. A complete ban on mobile systems could be monitored since the detection through satellite or other means of even a single mobile launcher would then be a violation. However, an upper limit on mobile launchers would be hard to verify because it would have been difficult to make an accurate count without very complicated agreed arrangements. Despite considerable U.S. pressure, our SALT I negotiators could not get any Soviet concessions on this score, and they were forced to fall back on a unilateral statement, not agreed to by the Soviet Union, that we would consider "the deployment of operational land-mobile ICBM launchers during the period of the Interim Agreement as inconsistent with the objectives of that agreement."

As the Soviets moved into a MIRV program of their own, however, and the future vulnerability of Minuteman could be more clearly foreseen, this U.S. position against mobile ICBMs started to erode. In 1974 Secretary Schlesinger initiated design studies for possible mobile ICBMs, and the various ideas brought forward in the Strat-X study five years earlier were resurrected. During the Ford administration mobile ICBM programs received increased attention, and in 1976 the administration was considering the deployment of the new MX missile on mobile launchers, each in a twenty-mile-long blast-resistant tunnel. The idea would be to move the missile back and forth in this hardened tunnel so that the Soviets would never know where the missile was at any given time. If they wished to attack a missile, they would then have to destroy twenty miles of the tunnel. President Ford's advisers hoped that this linear shell game would require the Soviets to expend a prohibitive number of warheads to destroy one of our ICBMs. This hope proved to be unfounded.

In its early days the Carter administration strongly opposed mobile systems. At this time the Soviets were in the process of developing their SS-16 ICBM, their only modern ICBM using solid rather than liquid fuels. It was feared that this would become a mobile system, particularly in light of the fact that the upper two stages of the missile were identical with the mobile intermediate-range missile, the SS-20. In a speech in February 1977 President Carter expressed the hope that both nations would exercise restraint in the deployment of mobile systems.

President Carter, Secretary Brown, and other senior Defense Department officials were all very skeptical about the Ford proposed tunnel system. They postponed the decision on the MX because they had serious doubts whether the tunnels could be hardened at sufficiently low cost as to make this an effective way of ensuring survivability of the Minuteman or any follow-on missile. Earlier the Congress had refused funds for the large-scale development of the MX missile until some secure basing scheme was available. Indeed later research experiments showed that the tunnels would provide little, if any, increased survivability; once the tunnels were breached at one point, the shock wave would travel down the tunnel and destroy the missile wherever it was. Studies were made to see whether blast-proof doors could be placed in the tunnel around the missile, but these proved impractical. The result was that the tunnel scheme had to be abandoned. It is fortunate that the basing decision had been deferred long enough to avoid wasting billions on this ineffective proposal.

The weapon systems designers were sent back to the drawing boards. One idea was to build a large number of vertical silos for each Minuteman or MX missile and shuttle the missile back and forth between them. This did not meet the verifiability requirements, however. No plausible way of ensuring that only one missile was in each cluster of silos or that extra missiles could not be moved in on short notice could be devised. This first version of the shell game was therefore rejected, but it is still not totally abandoned. The Reagan administration is giving it priority attention.

Making the missiles air mobile was another approach considered. In this scheme, one or two missiles would be carried from airfield to airfield in a large cargo plane and be capable of rapid launch shortly after arrival at the new site. Consideration was even given to launching the missile from the airplane itself, and an actual test launch with a Minuteman was carried out from a C-5A. These concepts were never popular, even in the air force, and in the winter of 1978 were rejected. Air mobility would have been inordinately expensive, the missiles less reliable, and the system probably less suitable for countersilo missions. Furthermore there were too limited a number of suitable airfields to prevent the missiles' becoming vulnerable to Soviet attack on them all. In addition civilian populations were uneasy at the notion of airplanes with nuclear missiles flying around above them or landing on airfields nearby.

Some perfunctory studies were made before 1979 of launching the missiles from small coastal submarines (Project SUM), but this did not have any support in the air force or even the navy. Another proposal was to put buoyant missiles on ships from which they could be distributed around the seas and launched while floating in the water (Project Hydra). By the end of 1978, despite many analyses, the Defense Department had not been able to come up with any agreed satisfactory solution for the deployment of a survivable MX missile. Even by June 1979, when President Carter announced the decision to proceed with the full-scale development of the MX, he was not in a position to announce the basing mode. None of the proposed solutions for the MX missile system completely satisfied any particular group, and none solved the basic security problems.

Yet political pressures inexorably pushed the program forward to the decision point. The basic justification for going ahead with the MX system was the contention that our Minuteman ICBMs were becoming vulnerable in the early to mid-1980s and that we could not afford to delay the decisions on this program. The selection of the missile to be deployed, however, was not related to the vulnerability problem. In fact by choosing the largest of the candidate designs, survivability was being made more difficult and certainly more expensive. The key criterion of survivability depended on how the missile would actually be deployed, but the decision on that issue was delayed beyond June 8.

The MX missile basing choice was not reached and announced until September 7, 1979. There had been wide differences of opinion, both within and outside the government, on the optimum scheme. Since none was truly satisfactory, the final selection was difficult.

The racetrack deployment mode with twenty-three shelters per missile together with an ability for the missile to dash from one shelter to another on warning of an attack was claimed to provide for the survivability the administration sought. SALT II would be an important contribution to this survivability, thus adding to the justification for ratifying the treaty. The arrangements for missile assembly outside the racetrack and for opening the shelters to demonstrate to Soviet satellite cameras that there was only a single missile in each cluster were claimed to satisfy SALT verification requirements. Furthermore the design did not contravene the SALT limits on launchers since unlike the multiple vertical launch point schemes, no one could logically claim that each horizontal shelter was a launcher. Thus by September 7, 1979, the president completed the decision-making process on the MX missile system, which he believed necessary for the ratification of the SALT II Treaty, which in turn was necessary to make the MX work. Unfortunately these decisions did not buy any votes for SALT. The deployment scheme did not provide survivability without SALT, and the characteristics of the missiles destroyed the basic objectives of SALT. The MX is a distressing example of a weapon system that undermined our fundamental security goals and itself in the process.

11 The MX Decision and the 1980 Election

The Carter administration's decisions in 1979 to proceed with full-scale development of the MX missile and to deploy these missiles in racetracks in Nevada and Utah did not end the debate on this controversial weapon. It only started a new round in much more public forums. Although these decisions were made to placate those people in and out of the government who always turned to weapons to solve security problems and to get support in the Senate for the SALT II Treaty, in fact they did not please anyone.

Predictably those already in favor of controlling strategic weapons found this to be a disastrous move in the wrong direction. They made this position clear to the president even before the June decision, but he ignored their advice. They believed that the MX negated all of the benefits that could be gained through SALT. With the MX program going forward with high priority, it became increasingly difficult to argue that the SALT II Treaty was a step, albeit a small one, toward bringing nuclear weapons under control. Rather it looked as if SALT II was the cover for raising the arms competition to a new and more dangerous level. The MX decision dampened the enthusiasm of many whom the president might have counted on to drum up support for SALT.

The weapons supporters also were dissatisfied. Although they were pleased that the missile itself was at last in full-scale development, they were very unhappy with the basing mode. They disliked the racetrack scheme, and some, such as General Daniel Graham, U.S.A. (Ret.), former director of the Defense Intelligence Agency, even testified that it had been proposed only to make it easier for President Carter to kill the MX program at a later date. Many objected to the characteristics of the system added to satisfy SALT verification criteria. These people, who were intrinsically opposed to SALT in any case, felt that these only magnified the complications of the system and made the costs less palatable. They preferred multiple vertical launch point basing, initially using the deployed Minuteman III, which would later be replaced by the MX missile in these same silos. They called the MX program too little and too slow to deal with a Soviet threat that they foresaw in the first half of the 1980s.

Meanwhile the SALT II Treaty ratification process, despite Carter's MX decisions, started to collapse. The fiasco created by the administration's and Senator Frank Church's mishandling of disclosure of the Soviet combat brigade in Cuba gave the SALT opponents the issue they needed in order to build opposition to any agreement with the Soviet Union. Senators who had been insisting on the MX as a bargaining chip in exchange for their SALT support merely turned their attention to a new hand in the weapons poker game and raised the ante. If SALT were to be ratified, the overall Defense Department budget would have to be increased, not by the three percent in uninflated dollars that Carter initially agreed to but five percent or even more. By going along with these increases and the MX missile in exchange for SALT, President Carter undercut a basic argument for the treaty: that by reducing the threat, it would supplant the need for new military programs.

At the same time as European leaders were insisting that a SALT II Treaty was a prerequisite for a decision to deploy the new long-range Pershing ballistic missiles and strategic cruise missiles in Europe, hawks at home were insisting that a decision to procure new European nuclear weapons was a prerequisite for SALT. After a vigorous debate, which had serious political repercussions in several European countries, and after much U.S. arm twisting, NATO decided on December 12, 1979, to go ahead with the procurement of these so-called Euro-strategic weapons. But SALT still remained in trouble. Finally in January, after the Soviet military aggression in Afghanistan, President Carter asked the Senate to postpone indefinitely the ratification process on the treaty. During the 1980 campaign President Reagan said that SALT II was fatally flawed, and negotiations should be started anew on SALT III. The SALT II Treaty as such was now dead.

This MX-SALT interaction is a striking example of President Carter's political failures resulting from his vacillations and tendency to compromise on critical issues. On the one hand he expressed strong support for strategic arms limitations, but then he persuaded the Soviet Union to leave a loophole in the treaty that permitted the United States to proceed with the MX program. Yet the MX missile and its basing mode undermined the essential principles of stable mutual deterrence, which were being sought in the arms limitation agreements. In the end President Carter did not get a SALT Treaty, and he was left with an MX missile system that neither he nor his secretary of defense had been enthusiastic about in the first place.

While the SALT Treaty was proceeding along its tortuous path, the MX missile system also began developing increased opposition, particularly in the regions where it would be deployed. Residents of Nevada and Utah, areas of the country normally prone to support all military programs, were initially willing to have the MX in their states, urging only that the local impact be minimized. But as public awareness of the system became more widespread, the realization grew that damage to their society and environment would be overwhelming, long lasting, and probably irreversible. Originally the air force wrapped the MX missile in the American flag and sold it to the community as vital to U.S. security, but as the debate continued, some of this security rhetoric became tarnished. The public became increasingly dissatisfied with the glib and often misleading presentations of the air force representatives, and the more questions that were asked, the more uneasy the people in the area became.

This public dissatisfaction rapidly spread to the political leaders. Originally Governors Robert List of Nevada and Scott

Matheson of Utah were favorably inclined toward the MX, but as time went on they became more and more skeptical. Congressional hearings were held in Nevada and in Washington. Defenders of the MX were not convincing in their justification. Air force proposals to bypass the legal restrictions on land acquisition and water rights still further inflamed public opposition and made political support for the MX more difficult.

In May 1980 Senators Paul Laxalt of Nevada and Jake Garn of Utah jointly organized hearings before the Defense and Military Construction Subcommittees of the Committee on Appropriations to investigate alternatives to the MX basing system. Secretary Brown and other Defense officials then announced changes in the design of the system to dampen criticism. Racetracks were no longer the preferred plan; instead the clusters would have the shelters in linear or straight road arrays. Other changes, such as separating the launcher from the transporter, were made in order to cut costs, which had been escalating from the estimates made only six months earlier.

Defense officials still supported the multiple protective shelter system in Utah and Nevada but, out of deference to Senators Laxalt and Garn, agreed to give serious consideration to split basing, which would move at least half of the system to other areas, such as west Texas and New Mexico. In approving fiscal year 1981 funds for the MX, the Congress stipulated that the secretary of defense must provide to Congress by February 1, 1981, an analysis of alternative locations for possible split basing. The air force environmental impact statement of December 18, 1980, fulfilled this requirement. This amendment and the impact statement were attempts to lessen opposition to local senators during the election. At that period both the Pentagon officials and the two senators were opposed to moving the MX missiles to sea in small submarines, a concept advocated by many nongovernment experts.

Nevertheless, despite rather widespread uneasiness about the MX, both the Senate and the House voted the funds requested for the proposed program. In the aftermath of the Soviet invasion of Afghanistan and in an election year, it was not to be expected that any major military program would be cut. Furthermore many normally skeptical congressmen did not want to oppose the MX publicly while SALT ratification was still a live issue. That there was already substantial opposition was a warning of a stormy future.

The MX missile also became the subject of some debate during the 1980 presidential campaign. At the Democratic convention R. P. Joe Smith of Portland, Oregon, a Carter delegate, led a move to amend the platform to delete language supporting the MX missile and replace it with words opposing its deployment. This was defeated only after a bitter fight in which Secretary Brown personally lobbied delegates on the floor and President Carter sent them handwritten notes. The Republican platform criticized President Carter for delaying the MX missile program and called for earliest possible deployment of the MX missile in a prudent, survivable configuration. The implication was that although they supported the MX, they were not in favor of the multiple protective shelter basing scheme. Candidate Reagan in his September debate with John Anderson said of the MX, "I am not in favor of the plan that is so costly . . . and will take thousands of square miles out of the West." He had earlier taken the position that "this proposed mode of deploying the MX [the racetracks] should be scrapped because it is unworkable." He reaffirmed this skepticism in March 1981.

After the election of President Reagan, the program that the new administration would adopt remained uncertain. There was little question, however, that the priority development of the MX missile would be continued. The new Secretary of Defense, Caspar Weinberger, kept the funding at the same level proposed by his predecessor. President Reagan's campaign security advisers tended to favor the multiple vertical silo approach as opposed to the horizontal protective shelters. They were not sympathetic with the arguments that this method of deployment would endanger any future verifiable SALT agreements, and they ignored the fact that the multiple vertical launch point scheme depended on the same shell game and did not provide survivability in the absence of limits on warheads in SALT II, which they opposed.

In addition, the new administration faced major practical problems in the proposals to build large numbers of vertical, blast-resistant silos and the necessary facilities and heavy-duty roads needed for them. Environmental and water level problems would be magnified. In the Great Basin of Nevada and Utah, only 60 percent of the proposed deployment area is suitable for vertical shelters because of the high water table. There could also be important delays in the program if a new type of mobile transporter-launcher system were to be designed, developed, and procured. Much of the experience with the horizontal mode would have to be learned all over again with a vertical one. The air force managers of the program reportedly were strongly opposed to any major changes in the design early in the new administration.

Many Reagan supporters have favored building ballistic missile defenses. The Republican platform called for creating "a strategic and civil defense which would protect the American people against nuclear war, at least as well as the Soviet population is protected." President Reagan in an interview during the campaign supported some form of defensive weapons. If strategic arms limitations reach a dead end, then the new administration may not believe the risks of losing the ABM Treaty very important. Certainly the new administration is looking seriously at ballistic missile defenses that could be built to overcome the vulnerability of all proposed MX multiple launch point schemes. Consideration was even being given to building an ABM to protect and defend the existing Minuteman silos, even though such a defense would be both impractical and expensive if every launcher holds a missile and no multiple launch point scheme is used.

There were also strong pressures within the Reagan administration to move at least an important part of the system out of Nevada and Utah. Senator Laxalt, President Reagan's closest adviser in the Senate, is quite sensitive to the local opposition in his state of Nevada. Other Reagan advisers called for using the existing Minuteman sites in Wyoming, Nebraska, Montana, and North and South Dakota. They argued that extra silos could be built for the existing Minuteman missiles, which could be replaced with the MX missile when it is available. But changing the direction of the program will not make the ICBM vulnerability problem go away and promises to incur many of the difficulties and delays that have afflicted the program in Nevada and Utah. Winter weather on the northern plains can make it very hard to move a missile from one launcher to another. Land acquisition can be even more difficult since in many cases valuable, privately owned farmland would be needed. The proposal to move part of the MX system to New Mexico and Texas has already started to generate local opposition in those communities. Senator John Tower, the new chairman of the Senate Armed Services Committee, is strongly opposed to any site in Texas, his home state. The governor of Wyoming earlier invited the MX into his state but subsequently took a public stand against MX deployment there. Realistically it is unlikely that any locality in the United States would welcome the MX system.

Changing the nature or design of the MX system is always an attractive option until thorough analyses have been made. But as detailed studies become available and local citizens become involved, the same weaknesses and problems tend to crop up over and over again. This has been the experience in the past as one scheme after another for building a survivable land-based ICBM system has been strongly touted as the best, only to be discarded a year later. The long underground tunnel scheme of the Ford administration is a classic example. The Reagan administration eventually may realize that there is no satisfactory solution for a survivable land-based missile system as long as there are no long-term restraints on Soviet strategic nuclear warheads.

III Security and the MX

12 Survivability

An evaluation of the decision to go ahead with the MX system deployment must weigh any improvement in security that it would provide against its costs in dollars, resources, and social and environmental disruption. If the system makes a truly vital contribution to the future survival of the nation, then major costs can be tolerated. But if the added security is nonexistent or marginal or if there are better alternatives, then important sacrifices should not be required of the people locally involved or of the nation's taxpayers. In some weapons decisions the balance between these two sets of factors is hard to determine—not so with the MX.

The central military justification for the MX system is to provide a survivable ICBM force. If the MX succeeded in eliminating a real vulnerability to an important component of our strategic deterrent, then it would have come a long way toward justifying itself. Therefore the survivability of the system should be evaluated with great care, for without it the system cannot be justified.

In order to determine the priority of this requirement, the vulnerability of the existing ICBM force and the viability of the strategic deterrent as a whole must be evaluated in realistic terms, not in the fantasy world of worst-case analysts. There is little quarrel that we must have a secure deterrent since this is the basis for avoiding nuclear war, and the consequences of a failure are so great that no reasonable risk should be allowed here. But a simple mathematical calculation based on a few simple assumptions can easily lead to an unrealistic assessment of the risk.

The best intelligence estimates indicate that at some time in the 1980s, the Soviet Union will have many thousands of high-yield warheads and that their accuracy, as determined by test launches, will also be high. Defense officials have reported that for the past two years the Soviets have been firing their ICBMs down their test ranges with guidance systems capable of directing their multiple warheads to within about 800 feet of their apparent aim points. If such accuracy can be obtained in an actual military operation, then each warhead, which could have yields of 600 to 1,000 kt, would have a high probability of destroying one of our Minuteman hardened silos. With about 1,000 silos in the entire U.S. ICBM force, the Soviet Union would need only about 2,000 such warheads to threaten this leg of our triad, even allowing for failures.

But the achievement of this accuracy in a real military operation could be quite different from demonstrating it in peacetime on a test range. Surely there must always remain a large uncertainty of success in the minds of Soviet leaders, who would have to make the decision to launch such an attack. Yet high confidence is a prerequisite for any head of state who might consider launching an aggressive attack when the consequences of failure are so great.

The ability of a warhead to come close to its intended aim point depends not only on dispersion from the guidance system but also on the gravitational forces along its trajectory and on the aerodynamic forces as it reenters the atmosphere. These latter two are not under the control of the weapons designer. Corrections based on satellite measurements can be made for gravitational anomalies in the Arctic regions, but there will always be some uncertainty because the polar trajectory followed in a real military situation is not one on which the missile warhead can ever be actually tested. Aiming errors produced by the wind forces during reentry can be reduced by designing the warheads to go through the atmosphere at very high speeds, but these errors can never be entirely eliminated. In an actual attack they can create for all warheads biases comparable to the errors in the guidance mechanisms. These aiming errors are not serious when attacking an easily destroyed industrial target or vulnerable aircraft on the ground; however, they could be quite important when the warhead must come close to a hard target such as a missile silo. In such a case the unknown errors could destroy confidence in the success of the attack.

A second factor that affects the reality of the threat to our Minuteman silos is the reliability of the Soviet missile system. Even under controlled test conditions, the Soviets reportedly are able to launch a single missile successfully only 80 percent of the time. It is not likely that this would be exceeded in an operational situation where large numbers of missiles must be fired at a specific time, set well in advance for political reasons, not technical preparedness. Launching a couple of thousand warheads successfully at a thousand targets on a very precise time schedule would be extremely hard to do, even under ideal circumstances. There are bound to be snags. A significant number of failures would certainly occur.

Correction for such failures is not simple. First the command center must obtain the information on which warheads failed and thus on which targets were not destroyed; then a second warhead must be reprogrammed to try again. Firing several warheads at a single target is not the easy solution because of the danger that later ones will be destroyed by the effects of earlier explosions. This so-called fratricide effect probably would not allow more than two warheads to be fired at any one silo within a short time interval. When all of these technical factors are taken into account along with human errors, it would seem unlikely that any Soviet leader could rely on destroying more than half of our 1,000 missiles in an actual attack. But even if 90 percent were destroyed, we would still have 100 ICBMs with 100 to 300 warheads with yields between 350 and 1,000 kt, certainly not an inconsequential force.

Probably the most important factor creating an aura of unreality to the threat is the lack of Soviet motivation to launch such an attack. Even were the Soviet leaders to be convinced that they could succeed in destroying most of our ICBMs, they would have to recognize that the United States could still retaliate with all its submarine missiles at sea and all its bombers on an alert status. Since the United States even in peacetime keeps about twenty-five of its submarines at sea at any one time, we could launch 400 missiles with about 4,000 warheads at Soviet targets. Although these warheads are not now as accurate as land-based missiles, they are still quite capable of destroying most military targets. All Soviet aircraft on the ground, military installations, armament depots, and the industry that supports the military establishment could be destroyed. In addition at least one hundred bombers on alert, each carrying twenty long-range, highly accurate cruise missiles to avoid the need for the planes to penetrate Soviet air defenses, could wreak havoc on Soviet military and civil targets. If the Soviets tried to destroy the bombers on the ground with a submarine missile attack, then this would be an unequivocal, timely warning, so that our ICBMs could be safely launched before the Soviet ICBM warheads could destroy them.

What could be the possible Soviet gain that would warrant running the risk of retaliation on this scale, even if their attack succeeded in knocking out all our ICBMs? The thesis that the United States would not respond out of fear that this would precipitate destruction of its cities would seem to be a slender reed on which to stake survival of the Soviet Union as a modern society if it proved wrong. No one can say what would be the presidential response in the aftermath of an attack against our ICBMs, but with millions of Americans killed and tens of thousands of square miles contaminated with radioactive fallout, the probability of U.S. retaliation would seem very significant. Fundamentally these scenarios of an aggressive Soviet attack against our ICBMs to obtain some political advantage or even of a blackmail threat to do so are unrealistic. No political leader would take such risks without extreme provocation. Only the fear that their deterrent was about to be destroyed by a U.S. first strike could justify launching an attack against our ICBMs.

Nevertheless it is an unsatisfactory situation to have a major component of our strategic deterrent triad vulnerable, even in paper calculations. Vulnerable forces become attractive targets, particularly if they are threatening, and produce incentives for initiating an attack to destroy them in times of extreme crisis. These incentives would be greatly increased if the warheads in our missiles had a capability of destroying a significant portion of the Soviet deterrent. Then they become flashing beacons drawing enemy fire. When the side that launches first stands to gain very significantly, the strategic balance is very unstable even though both have equal forces.

The new Minuteman warheads with their more advanced guidance system already have good countersilo characteristics; therefore when potentially vulnerable, these missiles could increase the risk of a nuclear conflict. Although the number of such Minuteman warheads (1,650) is too small to have a high probability of destroying the entire Soviet ICBM force, it is nevertheless big enough to arouse major concern within the Soviet hierarchy. Thus even though the chances of a Soviet first strike against our ICBMs, or even a peacetime threat to launch such an attack are very low, it is quite important to try to reduce this vulnerability insofar as this is feasible.

This objective does not require a hasty response. It would be better to do nothing than to take ill-advised, irremediable action when a simple way of reducing the threat to these Minuteman ICBMs would be to stop their modernization with threatening countersilo warheads, the only factor that can reasonably provoke an attack against them. Because the march of technology may be almost impossible to halt forever and because eventually even our Minuteman ICBMs will probably appear to the Soviets to be silo-busting weapons, the long-term answer may be their total elimination. This will be the ideal way of leaving Soviet countersilo weapons without any priority targets and rendering them less important.

Will the deployment of the MX missile in any of the proposed multiple launch point deployment schemes—the racetracks, the linear arrays, or the multiple vertical silos—actually provide the desired survivability? All of these concepts depend on the successful deception of the Soviet Union. The United States with its open society must be able to count on being the winning entrepreneur of a mammoth shell game in which there will only be a single opportunity to play. If through some inadvertence the Soviets are confident that they know which shelters hold the missiles, then the whole effort will have been in vain whether or not they were right. They could then decide to launch an attack. This seems like the wrong contest for Americans to be playing with the Russians. It is hard to believe that we can beat them in the deception game and even harder that we can ever be confident we have succeeded in fooling them.

The requirement to conceal from the Soviet satellites where the missile has been deposited on its travels around its cluster is probably the easiest to satisfy. There is no fundamental reason why with shields over the transporter, the missile and its launcher cannot be moved from launch point to launch point without a space camera's being able to detect where the missile has been left behind and when a dummy has been picked up. The opening of the ports in the tops of the shelters, as proposed by the Carter administration to permit verification, should not compromise this type of deception since only a small fraction of the ports need to be opened at any one time.

This is only the beginning of the deception game. More likely, little but telltale clues will give the game away. The missiles will have to be serviced to be kept in operational readiness at all times, and this requires people. Since the manpower requirements to have crews at all shelters at all times would become prohibitively expensive, the specific shelter that houses the missile cannot be staffed either. Major maintenance will require movement of the missile to a cluster maintenance center. But occasionally the shelter with the missile will have to be visited; other shelters will thus also have to be visited from time to time to avoid drawing attention to the shelter with the missile. The opportunities for mistakes will be great. Where people live and work day in and day out, year after year, the possibility of leaving identifiable patterns, which give the game away, will always exist. And we will never know whether this has been done.

This difficulty is compounded by the air force's stated plans not to maintain tight security within the deployment area. In an effort to make the land use by the MX more palatable to local inhabitants, security fencing is at present planned only around the hardened shelters themselves. The public—campers, ranchers, miners—are to be allowed free access to the roads on which the missiles will be moved and which the operational personnel will have to use for access to the shelters and the missile itself. These outsiders could bring sensitive instruments, dogs for smelling out explosives, or other aids for detecting where the actual missile is. Even private aircraft supposedly will be allowed to fly over the clusters.

This freedom of access is designed to improve the acceptability of the deployment to the local population, but in the end it seems almost certain that the air force will have to exercise very stringent security controls over the area. Probably these will have to be instituted gradually once the system is sufficiently far enough along that the MX deployment cannot be stopped. Originally in September 1979 the design did not call for dummies on the transporter when no missile is present; yet just six months later elaborate dummies with a weight equivalent to that of the missile and with simulators to send out similar infrared, nuclear, and electromagnetic signals were being planned. All of these may work, but in the long run it may be small, unforeseen things, which cannot be simulated by scientific techniques, that will lead to security breaches.

And this leads to what may be the greatest deception problem of all: How will the U.S. leaders know that they have successfully deceived the Soviet Union? Certainly the Russians are not going to tell them. We will always have to be concerned that there has been some human slip, some clue that has been let drop inadvertently. The United States can never be confident that it has succeeded in fooling the Russians. It will never know whether it has a truly survivable basing mode or whether the Soviet Union has beaten the shell game and knows which shelters to attack.

The deception problem would not be materially different were a decision made to move from horizontal mobility with hardened shelters to multiple launch points using vertical silos. Since the deception game has not been played through so thoroughly with the vertical system, it is not possible to say whether there are any obvious loopholes that could compromise the missile location from the start. It should be noted that the missiles will move more slowly from one launch point to another with the vertical system, therefore increasing the potential opportunities for observation. There will not even be a fallback dash capability on warning of an attack, so the vertical launch point scheme is more dependent on successful deception of the Russians. The security problems in protecting the area from ground observations will be similar to the horizontal systems unless the entire area, thousands of square miles, is cordoned off. The need for dummies will remain, and the mechanics of replacing a dummy with a missile in a vertical silo could be more complicated than in a horizontal case. Fundamentally, however, these vertical systems have the same deception problems, so there will be no way of being certain that the Soviets have not figured out which silo holds the missile.

In the long term it is likely that this uncertainty of successful deception will grow. Some people will proclaim that the Soviets have learned the secret of which shelter has the missile. No one will be able to prove them wrong. There will be demands for increased security over the deployment area, denial of access to the valleys, banning of aircraft overflights, and countless other restrictions on the local inhabitants. There will be a demand to forgo the special arrangements to facilitate verification. Total concealment of strategic systems will become the order of the day. This principle, if adopted by the Russians as well, could decrease our security with or without any arms control.

Finally it will be decided that deception is a losing cause. Instead of having only one missile in one shelter per cluster and dummies for the other, missiles will be procured for all shelters. This would be attractive since the biggest expense for the MX deployment will be the base structure. The total number of missiles could be increased twenty-three fold. A similar situation occurred in the past with MIRVs. When ABMs first created concerns that our warheads might not reach their targets, we incorporated penetration aids in our missiles to negate the missile defenses. One form of these was to include decoys that would look to Soviet radars and sensors just like warheads. In the end it was decided that the ideal decoy was another warhead, and this was the genesis of multiple warheads and then MIRVs. Soon we had 7,000 warheads instead of 1,700. History could repeat itself with the MX.

But even if the deception works, there is a more fundamental problem in ensuring survivability with any plan relying on the basic concept of any multiple launch point scheme. The effectiveness of such systems depends on the Soviet Union's not having significantly more warheads capable of destroying hardened targets than the United States has blast-resistant launch points. Thus the present 1,000-silo Minuteman system was not considered vulnerable when the Soviet Union had 1,400 ICBMs without MIRVs; however, when the Soviets began to put multiple warheads on their ICBMs and when these had the requisite accuracy and yield, then the Minuteman system became potentially vulnerable. The multiple protective shelter program called for increasing the number of potential ICBM launch points from

1,000 to 4,600. This number was considered large enough and was actually specifically selected on the basis that the Soviet Union would be allowed a maximum of about 6,500 ICBM warheads under the SALT II Treaty. Thus the Carter administration recognized that the effectiveness of the proposed basing scheme is directly tied to SALT. But we do not have a SALT II Treaty. The new Reagan administration has specifically said that the SALT Treaty is flawed and that instead we must move on to SALT III. Even under the best of circumstances, it seems unlikely that a SALT agreement putting upper limits on the number of Soviet ICBM warheads will be achieved in the near future. Even had the SALT II Treaty been ratified as the Carter administration hoped when it made the decision to go ahead with the proposed MX basing, that treaty would have expired in 1985, before the first MX missile was in its cluster.

A weakness in any multiple launch point scheme is that it provides little added survivability until the number of potential launch points approaches the number of Soviet warheads. Thus the MX system would have been of little value unless U.S. negotiators could have persuaded the Soviet Union to extend the life of the SALT II Treaty and kept the lid on Soviet warhead deployment indefinitely. The lot of a U.S. negotiator would not be enviable: he would have to convince his Soviet counterpart that the Soviet Union must extend SALT II and keep limits on the number of its own warheads in order to make the new threatening U.S. ICBM system from becoming obsolete before it was deployed.

If we cannot rely on SALT to limit the number of Soviet warheads, is there any method by which these multiple launch point schemes can be made survivable? This weakness of the system in the absence of limits on Soviet warheads was publicly acknowledged by Carter administration officials when the SALT II Treaty began to run into political difficulties. In testimony before the Senate in the spring of 1980 William Perry, under secretary of defense, said that the Soviet Union could have 3,000 countersilo ICBM warheads on its SS-18s alone as early as the end of 1983. The SS-19s could add significantly to this total. Without a SALT agreement there is no reason why they could not have 10,000 or 15,000 such warheads before 1990, when the MX with 4,600 shelters would become fully operational. Through 1981 the large Soviet SS-18 has been tested with only ten MIRVs, but our intelligence has warned that because of its large size, it could carry twenty or even thirty warheads. Their SS-19 similarly could increase the number of its MIRVs from six to twelve or more. Thus the Soviet Union would not have to procure additional missiles or design an entirely new missile in order to increase the number of its warheads by a major factor, making the 4,600-shelter MX basing obsolete.

The Carter administration's response to such a prospective buildup was to propose an increase in the size of the MX system beyond the planned 4,600 shelters. Figure 10 shows how the number of shelters needed to ensure that 100 MX missiles survive increases as the Soviets acquire more warheads. The number of shelters could be expanded to about 7,000 within the currently planned deployment sites. Further enlargement would require more land, and if necessary the number of missiles and clusters could be increased. A race between Soviet warheads and U.S. launch points would be on. Because the missiles and their potential launch points are all targetable, the size of the MX system is no longer finite; it is an open-ended project with no terminus. Recognition of this probable need for expansion in the light of the unconstrained Soviet threat probably prompted the change in the design of the MX basing from closed racetracks to more readily expandable linear systems. At the same time it was announcing the objective of reducing the acreage required for the MX deployment, the air force was actually planning for its

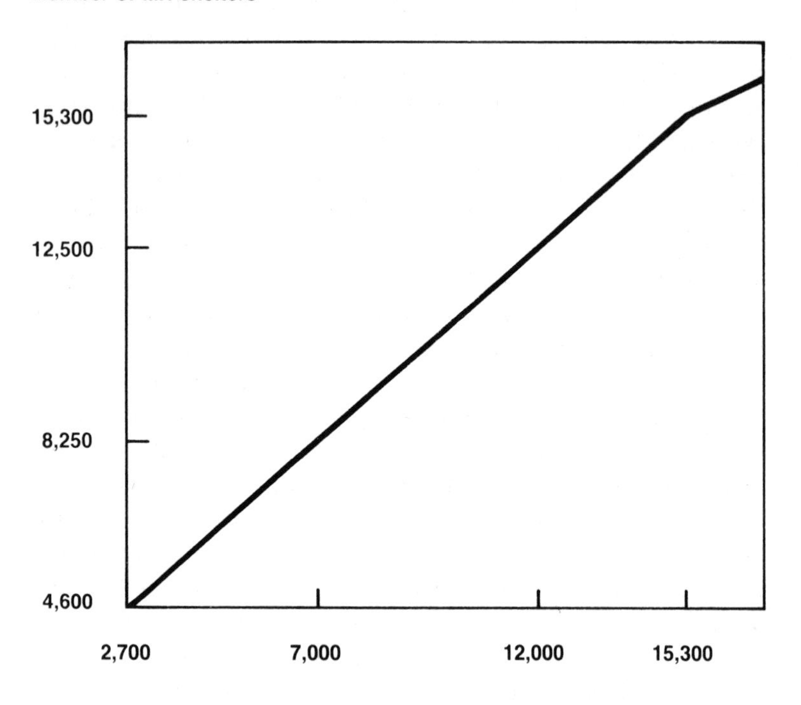

10. Multiple protective shelter requirement for survival of 100 MX missiles, if there is one MX for every 23 shelters and each Soviet warhead has a probability of 0.85 of destroying a shelter.

expansion. The projected costs can skyrocket still further. The desecration of the landscape will be even more extensive; no one knows where it will end.

The race to acquire more shelters than the Soviets can acquire warheads will be one in which the United States, but not the Soviet Union, will be running blindfolded. It will be relatively easy for the Russians to know how many launch points they need to have the capacity to attack since satellites can detect the early stages of construction. This information gives them a long lead time to take whatever countermeasures they may deem desirable. The reverse will not be the case for the United States; we will never be able to count the number of warheads that the Soviets have actually deployed. The SALT II Treaty provided verifiable upper limits on the number of warheads the Soviets could have since it was possible to observe from tests the maximum number of warheads each missile could carry and to count the number of each type of Soviet MIRVed missile launchers. Without SALT, this task will become very difficult, if not impossible. Now the Soviet Union is free to test an SS-18 with twenty or thirty warheads instead of ten. We could spot this test but would never know whether the missiles actually deployed had ten, twenty, or thirty warheads. While the yields would be lower, the warheads could still have about a 90 percent probability of destroying an MX shelter (table 3.1). Furthermore without SALT there would be no restrictions on concealing the nature of their tests, perhaps through encrypting their test data, so that we could not determine the characteristics of their missiles. In addition, if they chose, they could procure more than the 820 MIRVed ICBMs they are restricted to by SALT II.

Former national security adviser Brzezinski has claimed that "proceeding with this new system should improve Soviet incentives in SALT III by demonstrating our determination to maintain essential equivalence." He and other officials have claimed that when the Soviets see our determination to build launch points as fast as they build warheads, they will back out of the race. This is the same wishful thinking used to justify our MIRV deployments in 1970. Now Soviet MIRVs are threatening our Minuteman force. Because our MX missiles threaten their ICBMs, they will have strong incentives to produce enough warheads to threaten all of our shelters.

In any case, how will we know without SALT whether they have or have not backed out? Worst-case analysts will always assume that Soviet missiles are deployed with the maximum number of warheads that they have carried in tests. Thus we could move into a position in which we were racing ourselves. The Soviets might never add a single warhead beyond what they would have been allowed under SALT or perhaps even less, but we would be building shelters on the assumption that they were continuing the program. This is another reason why long-lasting arms limitation agreements are necessary, why all multiple launch point schemes are open-ended, and why we will never know how to bring these programs safely to an end.

It has been argued that we could build shelters faster and cheaper than the Soviets can build warheads. This is a dubious argument at best. The Soviet Union can have more than 6,000 warheads by 1984, two years before we have a single operational MX in its shelter. Building 4,600 shelters by 1990, our goal, will be a very difficult task, and we are falling behind every day. Certainly if the Russians can add 3,000 warheads to their stockpile between 1980 and 1984, they can add another 6,000 by 1990. Thus when the currently planned MX is complete, it will still be inadequate to meet the existing Soviet threat. We will always be in a losing race.

Comparative costs are difficult to estimate and in any case probably are not determining factors for either country once a decision is made to move ahead. It is not cheap, however, to build in remote areas hardened shelters with heavy-duty roads to connect them. Past experience with military construction does not provide the basis for confidence that costs will not escalate. There will always be tendencies to add refinements to the installations. The Soviets, on the other hand, can at least double the number of their warheads without major changes in their missile programs. They do not need to build new missiles or launchers; the only major cost would be for the nuclear warheads themselves. If they were determined to counter the threat of the U.S. MX missile—and they almost certainly will be because of its threatening nature—these costs would not deter them.

The MX system designers recognized that this race would have to end somewhere. If the number of Soviet warheads continued to grow, then they admit a new approach to making the MX survivable would have to be sought. Increasing attention is being given to a ballistic missile defense of the MX missiles in their launch clusters. The option to provide a meaningful defense of ICBMs was renounced by the United States and the Soviet Union in 1972 when the ABM Treaty was negotiated and ratified. Both nations recognized then that unconstrained ballistic missile defenses could create major uncertainties over whether the mutual deterrent posture persisted. As long as the Soviet Union had no ABMs, we could be confident that all warheads, whether launched from submarines or from land, would not be prevented from reaching their assigned targets. The Russians would know this too and would be deterred from aggressive action. Furthermore ABMs favor the aggressor because it would be easier for the fully alerted ABMs to cope with a more disorganized, less coordinated retaliatory attack. ABMs could significantly decrease the stability of the strategic balance and increase the likelihood of a nuclear strike without providing any real protection for any nation.

Now we are back considering the deployment of ballistic missile defense systems to protect our multibillion dollar investment in the MX missile. Theoretically an ABM system might be designed to improve significantly the survivability of the MX missiles. Defending targets where even a 50 percent failure rate would be quite satisfactory is a much easier task than defending a city where millions might be killed if a single weapon penetrated the defenses. Furthermore ABMs and multiple launch point ICBM-basing schemes can be synergistic in their protection. ABMs that rely on intercepting the ICBM warhead as it nears the ground (the army LOADS ABM is one) are more effective in protecting multiple launch points than are single silos. A single silo defense requires a complete ABM system for each silo, and it must launch an interceptor at every warhead attacking the silo. Such a contest will always give an advantage to the offense. It will be easy to overpower. Furthermore when the cost of the radars and other hardware is added in, the offense can quickly impoverish the defense.

With multiple launch points, however, the situation can be reversed. Because only the defender would know which shelter housed the actual missile, the ABM need only fire its interceptor missile selectively at those warheads that its radars see coming at a shelter actually housing an MX. This enables the defense to conserve its ammunition and keep from being saturated. Yet if the deception as to the location of the MX has been successful, the attacker must launch sufficient warheads at all shelters to overwhelm the defenses. Theoretically if it would take twenty-three Soviet warheads to be sure of destroying an MX missile in its shelter cluster, then even one effective ABM interceptor would force the Soviets to expend two warheads per shelter or a total of forty-six to be sure to destroy the missile. If it worked, this would shift the advantage to the defense; however, the attacker could adopt the tactic of firing one warhead at each shelter and then only firing additional warheads at the shelter he saw the ABM defend.

The logic of selective defense may work theoretically, but it is probably impossible to apply in an actual situation where the ABM itself could be a target. A fundamental requirement of any

ABM system for defense of missile sites is that all parts—radars, interceptors, and so on—must be at least as survivable as the missile they are defending. Otherwise the attacker merely first attacks the defense, and after neutralizing it attacks the ICBM. Inability to protect its radars was the Achilles heel of the Safeguard ABM defense of Minuteman sites in the early 1970s. Recognizing this necessity for ABM survivability, the army proposes to use deceptive basing for a mobile ABM system protecting the MX. The ABM radars, interceptors, and computers would all be concealed in an MX shelter (probably not the one containing the missile), moved from time to time from one shelter to another, and the shell game played with the ABM as well as with the missile. The ABM would have to be operated remotely, presumably from the command and control aircraft since any ground operations center could be knocked out. The maintenance required to keep the ABM ready at all times is much more complicated than that required for the missile, and yet this too would have to be done without disclosing to the Soviets where the ABM is hidden. The practical problems of deceiving the Russians will be extraordinary, and the risk will be great that the entire deception scheme for the missile will be compromised.

The most important weakness in this ABM solution, however, is not in its technical complexities but in the Pandora's box that renewed ABM deployment will reopen. The ABM Treaty would have to be abrogated or renegotiated. It is almost certain that under such conditions all types of ballistic missile defenses, not just those defending missile launchers, would end up by being allowed. It is naive to hope that we could negotiate a treaty allowing U.S. MX site defenses but not Soviet city defenses. The race would be on for large-area ABM defenses, which could significantly reduce the credibility of the deterrent. In an attempt to shore up the land-based leg of the strategic triad, we would have seriously degraded the reliability of both the land-

and sea-based legs. At the moment submarine missiles are the most secure part of our strategic force. With Soviet ABMs we would have renewed worries as to whether any of our missiles could retaliate effectively. Procuring ABMs to protect the MX or any land-based ICBMs would be an incredibly shortsighted policy.

In sum the proposed multiple protective shelter deployment of the MX missile will not ensure survivability of our land-based ICBMs even when fully deployed in the 1990s. This will be equally true of the multiple launch point scheme using vertical silos, which has been proposed by some of President Reagan's advisers. We will never know whether we have succeeded in deceiving the Soviets as to where the actual missile is hidden. We may succeed only in fooling ourselves into thinking we are safe. Without an assured long-term SALT Treaty limiting Soviet warheads, all multiple launch point schemes will not prevent the Soviets from threatening the destruction of the MX missiles. Instead this approach dooms us to a never-ending race between Soviet warheads and U.S. launch points, a race in which we will not only never know whether the Soviets are even running or whether we are winning. Attempting to shore up the survivability of MX by building ballistic missile defenses could compromise any hope of successful deception and in the long run will only reduce the reliability of both the land- and submarine-based ballistic missile legs of the deterrent triad. Multiple launch point basing of any type is fundamentally flawed. It is only a bottomless hole into which to pour our national resources.

13 The Risk of Nuclear War

When the decision to go ahead with the MX was made, Carter's national security adviser, Zbigniew Brzezinski, asserted that the improved strategic power of the MX missile will "reduce Soviet incentives to initiate an attack against our forces by giving us an ability to respond in kind." Thus the MX missile with its 2,000 countersilo warheads was justified because its 1,000 surviving warheads would discourage the Soviets from launching their countersilo warheads at our missiles. Matching the Soviets in threats against strategic deterrent forces seems to have been a basic goal of the MX, but this logic does not hold up.

Why should the MX threat to the Soviet's ICBMs reduce the likelihood that the Soviets would use their missiles against ours? Common sense tells us that just the reverse would be the case. If the Soviet Union knew that a U.S. attack could destroy a large fraction of their missiles, wouldn't that increase rather than decrease their incentives to launch first, before we destroy their missiles? Why should a threat to their ICBMs make them more reluctant to initiate an attack against us? In times of high tension the Soviet leaders would be much more inclined to attack if they feared that a delay would cost them a large part of their forces. Then Secretary of Defense Harold Brown said that our strategy should be not to give the Soviet Union any incentives to initiate a nuclear exchange, but the deployment of silo-busting weapons, which Brown has defended, appears to run directly counter to this strategy.

The Soviet Union can view the MX deployment only as evidence that the United States is seeking a first-strike capability. In March 1977, when defending the initial Carter SALT proposal, Brzezinski said that the Soviets should recognize that the MX system "could be extremely extremely threatening to them" with its "first strike capability against their land-based system." Later in 1979 in defending the MX, however, the Carter administration protested that we sought only the ability "to respond in kind." These declarations are hardly likely to reassure Soviet military leaders. When national survival is at stake, we do not put our trust on statements by Soviet leaders. Can we expect them to react differently?

Particularly in the case of the MX our actions must speak louder to the Soviets than words. A countersilo capability is effective only if it is used in a first strike. Silo-busting warheads used in retaliation will crush only silos, not missiles. The Soviets will have emptied some of their silos in their first strike, and they will surely empty the rest when they detect our retaliation underway. Certainly we are not so naive as to expect the Soviet Union to leave any missiles not launched in their initial attack in their silos to be destroyed by our second strike. Soviet radars or satellites would tell them that our countersilo missiles had been launched; in the fifteen minutes warning, they would get every missile out of their launchers that they could. Such launch-on-warning is a risky procedure in peacetime but would be a prudent Soviet strategy after they had started a nuclear war and killed millions of Americans. In fact the best way of ensuring that the Soviets launch any reserve missiles left over after their first strike would be for the United States to attack their ICBM sites in a second strike. It is an invitation for them to empty their silos and hit every possible target in the United States.

This is just the opposite of the scenario most often used to justify the acquisition of an MX with a countersilo capability. That scenario envisions a Soviet attack against our ICBMs and argues that unless we could destroy Soviet silos in response, any threat of retaliation would be incredible because we would fear this would lead only to Soviet destruction of our cities by their reserve strategic forces. Thus we would have no alternative but to bow to Soviet wishes.

The logic of this scenario is not sound. The surest way to get the Soviets to attack our cities would be to threaten the destruction of their missiles. Then they would have no choice but to use them against any target they could find—our cities, our people, our industry. Second-strike countersilo missiles are not a deterrent because the Soviets would understand the futility of such a response. They would be much less likely to be deterred if they knew we would waste our warheads on empty missile silos instead of on other military-industrial targets.

This point is also ironic; the Soviets would never consider such an aggressive first strike unless they were very confident that they could destroy virtually all our ICBMs. In such a situation it would not make any difference whether the ICBMs had countersilo capabilities or not since none would be left. If they failed in this goal, then certainly they would prefer to see our few surviving warheads aimed at silos, not airfields, antiaircraft defenses, shipyards, or military depots. Therefore the characteristics of the missile itself are not a deterrent to a Soviet decision to start or threaten nuclear aggression. In fact if the missile has countersilo characteristics, it has the opposite effect of increasing incentives to launch a nuclear strike.

Although the logic that foresees a suicidal Soviet strike at our ICBM sites is convoluted at best, the consequences of a U.S. first-strike capability on Soviet ICBMs are much more straightforward. Seventy-five percent of their deterrent lies in their fixed land-based ICBMs. Furthermore their submarine missile forces are much more vulnerable than ours, and their intercontinental bombers are very old, outdated, and vulnerable if caught on the ground. They cannot view such a threat with equanimity even though they should realize that an actual U.S. attack is not credible. If the United States were willing to consider seriously deploying the MX missile system at an ultimate cost of $100 billion even though its overall deterrent is not actually threatened, then the United States should be prepared for a very strong Soviet reaction to the threat of its MX.

Unfortunately all the actions they might take would be counter to U.S. security. The simplest and also the most reliable measure to ensure that their missiles are not destroyed by a U.S.

first strike would be to put their ICBM systems on a hair-trigger alert in peacetime. They would prepare themselves to launch these ICBMs when their early warning radars or satellites obtained evidence of a U.S. attack. Technically this should not be difficult for the Soviet Union even though their readiness is believed to be inferior to that of the United States.

But Soviet reliance on launch-on-warning would be extremely dangerous for us. It would short-circuit their high-level command and control and make our security dependent on the absence of false alarms in their warning systems. It is not a reassuring situation to have our survival dependent on the reliability of Soviet computers. The chances that we would be devastated in an accidental nuclear war would rise markedly if our MX drives them to adopt this tactic. This eventuality is far more likely than the possibility that the Soviets might launch a first strike against our ICBMs in the absence of the MX. The MX is substituting a high risk for a low one.

A second response that Soviet leaders might take when faced with the MX threat to their ICBMs would be to plan to launch a preemptive strike in time of crisis. Rather than wait for the United States to initiate the use of nuclear weapons and destroy a large fraction of their ICBMs in their silos, they would be under great pressure to fire first. Consider a conflict in Europe that threatened to escalate to the nuclear level. In 1975, Secretary of Defense James Schlesinger said that in the event of a Soviet aggression in Europe, we must be prepared to launch a first strike against their strategic forces. With the MX in existence, Soviet leaders would have strong incentives to eliminate as many of our strategic weapons as possible before we put the Schlesinger threat into effect.

Neither of these two tactics, launch-on-warning or preemptive strike, would involve major expense on the part of the Soviet Union. They can be implemented rapidly, primarily by political decisions. But both bring nuclear war closer.

In addition the Soviets could try to reduce the vulnerability of their ICBMs in much the same way that we sought to reduce the vulnerability of our missiles through mobility and multiple launch point basing. This was not an easy solution for the United States (in fact we have not yet solved the problem); however, technically it might be even more difficult for the Soviet Union even though deception would be easier for them. Virtually all of their ICBMs are liquid fueled, and such missiles cannot be transported when loaded with fuel. To move them from one launch point to another, first the fuel would have to be off-loaded, next the missile moved slowly and carefully to the second site, and finally it would have to be refueled and prepared for firing. Such an operation would require the minimum of a day and probably much longer. Furthermore vertical silos, not horizontal shelters, would be required since the liquid-fueled missile cannot be raised from a horizontal position to be fired. In the future they may move to solid fuels, but so far they have been backward with this technology.

It is quite likely that any mobile basing scheme that the Soviet Union would come up with would not be verifiable by our rigid standards. With or without any SALT agreements we would be less able to predict the size and nature of the Soviet threat. Our concerns over Soviet buildups would be magnified, not reduced, by the MX. In sum all of the potential Soviet reactions to our MX threat to their ICBMs would be detrimental to our security. We would be moving into a much more hazardous position and would be much more likely to be subjected to a never-ending arms race and to nuclear war.

In response to criticisms of the MX's countersilo characteristics, proponents have claimed that the ability to destroy blast-resistant targets is needed to satisfy other military objectives. We must be able to knock out nuclear storage sites, submarine pens, and underground command centers. President Carter's directive, PD 59, for preparing to fight limited nuclear wars specifically earmarked for destruction underground shelters housing Soviet leaders who had control over their nuclear weapons. This would not be a logical objective if we were serious about trying to keep the war limited. But even if it were, we hardly need the MX system with 2,000 new hard-target warheads to threaten all the Soviet command posts, nuclear storage sites, and other blast-resistant targets. The existing Minuteman missiles with 1,000 to 1,500 accurate high-yield warheads are more than enough. Furthermore we are proposing to equip in the next few years (long before the MX is available), 120 bombers to carry twenty highly accurate, long-range cruise missiles apiece. These can destroy very hard targets five to ten hours after takeoff, seemingly a satisfactory alternative to the MX since most hard targets do not have to be destroyed in the first hour of a war. Nuclear storage sites do not move away. The addition of 2,000 new, hard-target MX warheads can be explained only on the basis of a desire for a silo-busting capability.

Although the dangers to the stability of the strategic balance of a countersilo MX force have been widely recognized, nevertheless claims are still put forth that we must have such weapons in order to match the Soviet Union. Even before the Russians began deploying MIRVs on their larger ICBMs, predictions were made that they would have a countersilo capability in the 1980s. Matching the Soviet Union in this future weapons capability was a position supported by most politicians. They feared that the failure of the United States to acquire an equivalent countersilo capacity would be perceived as a sign of weakness, which could lead to political vulnerability. This perception probably would not have arisen if we had not advertised this as a weakness and ignored our strengths. In fact the criterion of counterforce capability was fairly exotic when it was first raised by Schlesinger and probably never would have been widely considered a measure of strength had it not been made an issue deliberately. In the years that the United States had a superiority in counterforce, no one has ever taken advantage of this situation or thought to use it for political gain.

Procuring forces just to match the Soviets is a mindless path to security; it allows the Soviet Union to make the decisions on the direction of U.S. programs. It ignores the military consequences of these decisions. Matching the Soviets in countersilo capability does not reduce the dangers from that threat; it only makes it more real. If the missiles in our silos can knock out Soviet ICBMs, then a first-strike countersilo capability would be necessary for the Soviet military. Then their missiles would have a military purpose; it would provide them increased incentives to use their missiles and to fire them first.

The MX missile cannot be justified as a damage-limiting weapon; just the opposite, it can increase the damage. Because of its counterforce characteristics the missiles become a magnet for ever larger numbers of Soviet warheads in the event of any actual conflict. In order to destroy hardened missile silos, warheads must be exploded close to the earth's surface, conditions that maximize radioactive fallout. Therefore any countersilo mission can lead to radioactive contamination over very wide areas. There will be no such thing as a surgical nuclear war between opposing missile forces, leaving the nations free to continue to exist as before.

If the Soviets were to attack our existing 1,000 Minuteman sites, 2,000 one-megaton warheads would probably be required. The casualties from such an attack have been variously estimated as between 2 million and 20 million people. The size of the area made unlivable would be very large. Fortunately most Minuteman sites, with the exception of those in Missouri, are relatively far from populated areas.

But were the MX, with a total of 4,600 shelters, to be deployed in Utah and Nevada, then the Soviet Union would have to launch about 9,000 warheads in order to have a reasonable hope of knocking out most of our 200 MX missiles. (This, of course, assumes that we have succeeded in deceiving them as to which shelter hides the MX.) In reality the number of warheads used by the Soviet Union probably would be considerably greater since we are already considering doubling the number of shelters. If the MX deployment occurred entirely in Nevada and Utah, then the area contaminated by radioactive fallout would in normal weather extend eastward across the Rockies to Denver and the farmlands of the Middle West. The heart of our farmbelt would probably be sufficiently contaminated to prohibit sowing crops for safe human consumption for several years thereafter. If the MX system were partially deployed in New Mexico and western Texas, then the radioactive contamination would most likely also spread eastward throughout Texas and Oklahoma and on into the southeast. But the West Coast cannot be considered immune and safe. Winds do not always blow from west to east; in the Mount St. Helens eruptions the winds carried volcanic ash to Portland and Seattle. In figure 11 are shown the contours for typical radiation exposures after a Soviet attack comprising a single 1 mt warhead on each of the 4,600 MX shelters in Nevada and Utah. This is the smallest possible attack to have any hope of neutralizing the MX system. If 9,000 warheads were used, the radiation exposures would be doubled. If the Soviets simultaneously attacked other strategic bases, the contamination would be spread over most of the United States.

One 15 mt explosion contaminated approximately 7,000 square miles with lethal fallout in 1954. The size of the area contaminated following an attack would be roughly proportional to the total megatons involved. In an attack against the MX system 10,000 mt could well be employed, and millions of square miles might be subjected to lethal fallout. General Lew Allen, chief of staff of the air force, in an unguarded moment stated that MX deployment areas in the Utah and Nevada deserts "could be a great sponge" to absorb a large number of

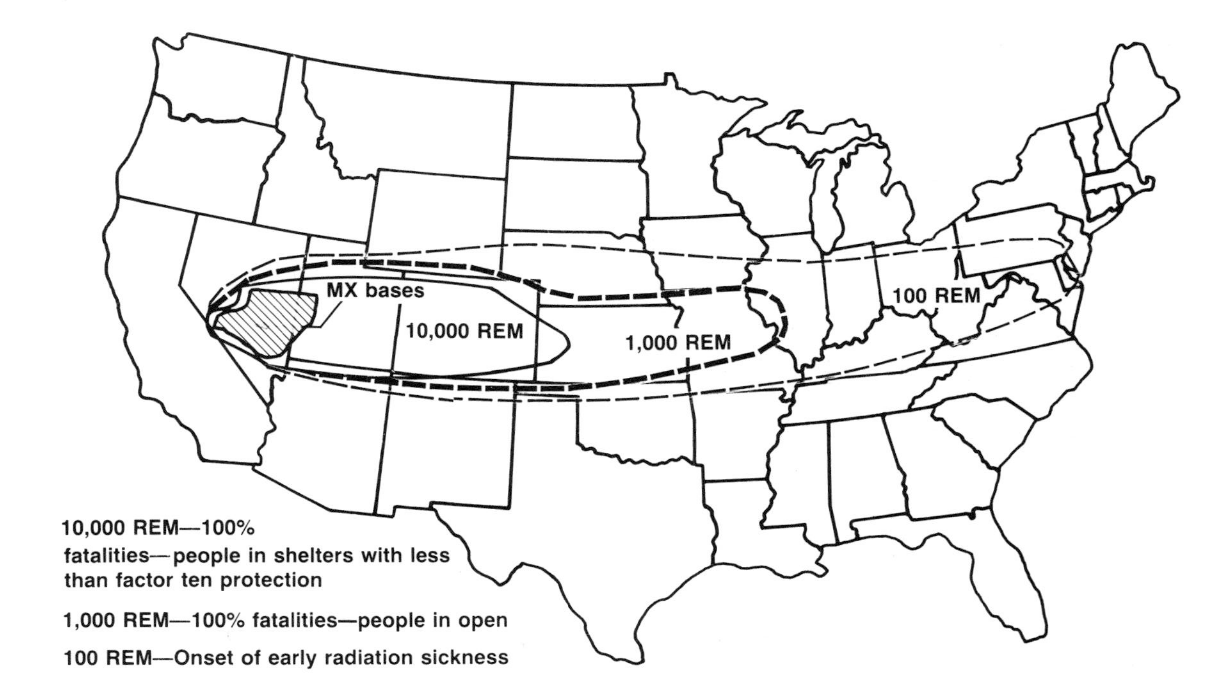

11. Radiation exposures following minimal Soviet attack on MX bases (4,600 one-megaton warheads, typical wind pattern).

Soviet missiles in an actual nuclear war. Unfortunately this radioactive sponge could be a large part of the United States. Building large numbers of launch points for missiles that threaten the Soviet deterrent is not a damage-limiting measure; it is only an invitation to even greater horror.

The MX missile system is an invitation to a nuclear war, engraved in U.S. gold. By its threat to the critical ICBM portion of the Soviet strategic deterrent, it will increase the likelihood of a nuclear conflict either by accident or through a preemptive strike in a time of crisis. It will decrease, not increase, the credibility of the U.S. deterrent. Seeking to achieve an ephemeral perception of equality with the Soviet Union by having the MX match its projected countersilo missiles is a dangerous prescription for destroying the stability of the strategic balance in the nuclear age.

14 Implications for Nuclear Arms Control

The deployment of the MX missile system with its 2,000 countersilo warheads is bound to create pressures for escalating the arms race and for making the future control of nuclear weapons more difficult. History has shown that the deployment of a major new weapons system by one country inevitably brings demands for the deployment of new weapons by the other. Therefore it is logical to expect a Soviet response to the MX with its missiles shuttling between 4,600 launch points. The Soviet Union probably will increase the number of warheads their SS-18s and SS-19s carry by factors of two or more in order to maintain a threat to our 4,600 shelters. Pentagon officials have already said that in such circumstances they would propose to double or triple the number of launch points and perhaps even deploy more MX missiles or even nuclear-tipped antiballistic missiles. And so, on with the arms race . . .

SALT

These pressures for arms buildups will make it more difficult to limit strategic arms. The SALT II Treaty did not directly place any real limitations on the MX program, although under the Protocol the testing of the MX in a mobile mode would have been temporarily banned. Because the Protocol would have expired at the end of 1981 and because the MX missile would not be ready for such tests by that time, this restriction would have had little significance. The decision to proceed with the mobile MX in its multiple protective shelters, however, has made it almost certain that these provisions of the protocol will not be extended beyond 1981 even if any other agreements are reached. And even that is now very much in doubt.

The SALT II Treaty's upper limits of ten warheads on any new ICBM and on the large Soviet SS-18 ICBM and of six on their somewhat smaller SS-19 were essential to the survivability of the proposed MX system or any other multiple launch point basing. If the Reagan administration tries to negotiate a so-called SALT III Treaty in place of SALT II, these limits on warheads per missile probably will disappear. At the very least the Soviets will require some major concession if they are to be persuaded to maintain a cap on the number of warheads on their large missiles. Similar problems may also occur with other current treaty provisions, such as the ceilings on the number of MIRVed ICBMs. In fact if the MX program is continued on its present path, this could lead to the unraveling of the balanced framework agreed to in the now-moribund SALT II Treaty. The MX deployment, whether in multiple horizontal shelters or vertical silos, may preclude any strategic arms limitations.

ABM Treaty

The most important agreement between the United States and the Soviet Union on limiting nuclear weapons that has been achieved to date is the ABM Treaty of 1972. Until recently pressure to modify or scrap this treaty has been relatively minor because its value to our national security has been quite generally recognized. Although this treaty is of unlimited duration, the two countries review its workings every five years to determine whether it should be amended or ended. The first review in 1977 was relatively perfunctory, and no attempts were made to suggest any changes. The second one in 1982 may not proceed as smoothly.

In the interim the MX has been authorized, and it has become increasingly apparent that multiple launch point basing does not by itself ensure ICBM survivability. In the absence of firm limitations on Soviet warheads, demands are growing that we consider building ABMs to defend the MX and thus shore up its survivability. Multiple launch point basing makes it technically feasible to build a system that might give some advantage to the defense, since the ABM interceptor need only be fired at warheads attacking a shelter actually hiding an ICBM. However, it ignores all the practical problems of making the ABM system itself survivable.

Since such an ABM cannot be built under the ABM Treaty, the MX will provide new impetus to the moves to scrap that treaty. These may not, however, reach the critical point by the time of the review conference in 1982, because the United States probably would not then be in a position to make a decision to deploy a new ABM system. But sometime in the 1980s, if the MX deployment continues, this demand for the termination of the treaty could increase. As billions upon billions of dollars continue to be spent on the MX deployment and as it becomes increasingly apparent that ICBM survivability is not being obtained, then the pressures to try to justify the existing investment could mount significantly. That building an ABM will only be wasting more billions could well be ignored even though an end to ABM limitations and the beginning of an ABM race could decrease the reliability of both the land- and sea-based legs of the strategic deterrent.

Comprehensive Test Ban Treaty

The achievement of a comprehensive test ban treaty to end all nuclear tests has been a worldwide arms control goal since the beginning of the nuclear age in 1945. Nevertheless, its successful negotiation is still in the future. One of the stated reasons why the Defense Department and the nuclear weapons laboratories have been reluctant to accept a total test ban has been the desire to complete development of an improved warhead for the MX missile. This requirement is not militarily significant, however, since the MX missile could be deployed with existing warheads. In fact the first model MX is designed to have a warhead identical to the one already completely tested and now being deployed in the Minuteman III. Nevertheless arguments can and will always be made that we must have even more efficient and larger yield warheads in the future.

Similar debates may be going on in the Soviet Union between its political leaders and its military and weapons developers. At the present time the Soviets probably have a satisfactory warhead for their existing SS-18 and SS-19 MIRVed missiles, but if they wished to double or triple the number of warheads to counter the MX, then they could easily have a requirement to continue testing until an optimum, light, but high-yield warhead is available. This can be quite critical to them if they wish to maintain a countersilo capability for their warheads, as they increase the number per missile. Thus both the United States and the Soviet Union may be reluctant to move to a ban on all nuclear tests if the MX missile system goes forward. This is another example of where the failure to move toward arms control reduces our security.

Cut-Off of Fissionable Material Production

Another arms control measure, which has received little attention in recent years, is a cutoff on the production of fissionable material (plutonium and highly enriched uranium) for weapons. This was a measure advocated by the United States in the 1950s and 1960s but never given sufficiently serious consideration by the Soviet Union to get detailed negotiations underway. In the last decade the United States has not renewed its initiatives on this measure. In September 1980 President Carter, in direct contrast to President Johnson, who moved the other way, made a decision to initiate programs to expand, not reduce, the production of fissionable material for weapons. The need for material for MX warheads was one of the points made to justify this decision.

To the Soviet Union, the MX makes a cutoff agreement even more unpalatable than in the past. Soviet material stockpiles are believed to be smaller than those of the United States, and if they are to enter into the anticipated race between Soviet warheads and American MX launch points, then their requirements could mount significantly. Not only must the number of warheads increase, but as the warhead size is reduced in order to pack more warheads into a given missile, more fissionable material is needed per warhead. Thus although the prospects for a cutoff of fissionable material for weapons are not good under any circumstances, they become even more dismal as the MX program proceeds.

Verification

The most specific interaction between the MX missile system and arms control lies in the area of verification. Without means of detecting violations of an agreement to provide confidence that potential cheating cannot jeopardize security, arms control would soon reach a dead end. Although arms control can promote mutual understanding, it must not be dependent on trusting the Russians. The ups and downs of international political relationships are too frequent and too varied to expect a treaty based on good faith to survive for long.

In almost all situations verification must depend on national technical means—established treaty language for technical intelligence methods that are now internationally accepted, such as satellite cameras. Sending inspectors into another country to check on compliance, so-called on-site inspections, in some situations can supplement national technical means, but such methods of getting evidence infringe on national sovereignty, are rarely useful, and are always difficult to negotiate, particularly with the Soviet Union and its closed society. Therefore an effort is made in arms control negotiations to frame the limitations in such a way as to make the provisions verifiable by national technical means. In some cases provisions are added to make these more effective. Interference with national technical means has been forbidden, and attempts to conceal activities related to the provisions of the treaty are also banned. Verification was dealt with very successfully in SALT I and SALT II. Without the restrictions of SALT II our knowledge of the Soviet threat, so important to our security, can be seriously degraded. The MX can also create grave problems for ensuring treaty compliance in the future.

The most important national technical means of verifying strategic arms control agreements are photographic satellites. Two types of camera systems commonly are used for this purpose. The first is for a surveillance-type mission to get wide area, moderate resolution coverage in a relatively short period of time. The entire Soviet Union can be photographed by such a system in a single day, but in practice it is somewhat better to phase this over several days. Such cameras are capable of detecting Soviet missile launchers and the associated support facilities long before they are operational. They can also see any missiles that are exposed but not those hidden in a covered shelter or silo. This is why the MX shelters have ports that can be opened when a Soviet satellite is passing by.

A second camera system is designed to zoom in on a previously detected target in order to obtain high resolution photography for detailed identification and measurement. Thus it is quite feasible to differentiate between a launch complex for the large Soviet SS-18 missile and one for its somewhat smaller SS-19. The lateral dimensions can be measured quite accurately, probably within a foot, so that any significant change in its maximum size can be determined. Roads and even footpaths are easily detectable so that secret activity in remote areas would run a high risk of detection.

In the case of the MX deployment in Utah and Nevada, it would be very difficult to sneak extra missiles into a launch cluster even at night without leaving evidence for the camera to find in daylight. The barriers placed on the access roads probably could not be breached without leaving some visible traces of activity, which would raise suspicions. Similarly moving a missile into a cluster by circumventing the barriers would be likely to leave evidence behind. This is particularly true in Nevada and Utah where any disturbance to the fragile, untouched terrain is easily observed from satellites.

On the other hand fooling the Soviet satellites as to the location of the missile may not be easy. For example, we have decided we need dummies to deceive the Soviets on which shelter or launch point hides the missile. But this could make it more difficult to provide convincing evidence to Soviet satellite cameras that the shelter or silo is empty. False alarms seem likely. Could the Soviets satisfy us if they adopted similar procedures? Concealing an ABM in the shelter would make this problem even worse. Thus the practical obstacles in ensuring that the Soviets cannot determine where the missile is hidden and that the shell game is not penetrated in the end could compromise the ability to verify with confidence that no more than one missile in each cluster is deployed. It seems inevitable that there will always be a dichotomy between the requirement to fool satellites on the location of the missile and the requirement to verify that only one missile is deployed in each cluster. When we want them to know that a shelter is empty, they must be able to know; when we do not want them to know, then they must not know.

With verification very much in mind, the Carter administration designed the racetrack basing and later the linear arrays with special characteristics. The ports in the tops of the shelters, which could be opened when Soviet satellite cameras were overhead, are one example. The government probably could support the claim that the U.S. MX basing was verifiable and that no action was being taken to prevent an accurate Soviet count of the number of missiles deployed. If the Reagan administration were to go back to the multiple vertical silos, new problems might develop. Barriers on the access road might work equally well for this type of system; however, providing reassurance that only one silo in a complex contained a missile might be more difficult. Removal of the cap on the silo as a satellite passed over might not provide the same degree of confidence that no missile was inside. A satellite camera might have more trouble in looking down into a vertical silo and being assured that it was empty or that a dummy was not a missile than it would to see into a horizontal shelter. Placement of some reflector or camouflage on the top of the hole to make it look empty should be relatively easy. Verifiability was one of the reasons for selecting horizontal shelters instead of vertical silos in the first place.

The vertical silos have one other SALT-related difficulty. In this scheme the missile would be moved vertically in its launch canister from one silo to the other. Under the terms of the SALT Agreement, the ceilings are placed on launchers, not missiles, because of the difficulty in counting the total number of missiles produced. Supporters of the multiple vertical silo approach claim that this scheme would not involve a proliferation of launchers since the missile is launched from the canister and there would be only a single canister per missile. However, this somewhat contrived interpretation that the canister and not the silo is the launcher could be the subject of objection by the Soviet Union. Although there is no official definition of a launcher in any draft treaty, in the past this has usually been taken to mean silos. Furthermore, submarine missiles are stored in canisters, which are not counted as extra launchers. Thus it is possible that because of the verification difficulties and the definition of a launcher, the deployment of multiple vertical silos can throw a serious roadblock into future SALT negotiations.

The problem for the United States is not whether the Soviets can verify the number of missiles we have deployed; it is to be certain that the Soviet Union is not deploying significantly larger numbers than are allowed under any agreement. It is very desirable for the United States to set good precedents for verifiability because we do not wish to give the Soviet Union any excuse for eroding the verification procedures, which are so much to our advantage. But in this case, the Soviets may not be able to deploy verifiable mobile ICBM systems. They may not be able to adopt horizontally mobile systems with shelters that conceal the location of the missile and yet can be opened for inspection by satellite cameras. They may be forced to adopt the nonverifiable vertical multiple launch point scheme. But once we have set the precedent of mobile missiles with multiple launch points, they might follow the same principle but with a different, unverifiable system. Then we would be unable to know how many missiles the Soviets have deployed. This would not only foreclose the achievement of a ceiling on ballistic missiles, but it could also downgrade very significantly our knowledge of the Soviet threat even in the absence of a treaty.

Furthermore under SALT II, no prohibitions were placed on the production of excess missiles. Even now when the treaty limits the number of launchers, some critics have worried that the Soviets would stockpile missiles, and then after firing the first salvo, they would reload the existing launchers. This is not a practical maneuver for current Soviet missiles, but this fear would be compounded if there were large numbers of excess launchers already built and the stockpiled missiles could be moved in and launched on short notice after abrogation of the treaty. The multiple vertical silos make evasion of the treaty ceiling by this method very easy and perhaps not very difficult even with horizontal shelters. Any multiple launch point scheme is certain to increase the uncertainty as to the nature and size of the Soviet threat with or without a treaty.

Nonproliferation of Nuclear Weapons

Finally the MX will have important implications for our security goal of trying to halt the spread of nuclear weapons to additional nations. Under the Non-Proliferation Treaty signed in 1968, both the United States and the Soviet Union "undertook to pursue negotiation in good faith on effective measures relating to cessation of the nuclear arms race at an early date and to nuclear disarmament." With the exception of the ABM Treaty, no progress has been made in meeting this obligation since the Non-Proliferation Treaty came into force. At the 1980 review conference of that treaty, both nations were subjected to strong criticism for their failure to live up to their commitments. The decision to proceed with the MX and its many additional MIRV warheads is viewed worldwide as a demonstration of lack of U.S. sincerity in its efforts to control nuclear weapons. The difficulties that the MX will create for achieving most other nuclear arms control measures are likely to prolong almost indefinitely any demonstration by the nuclear giants that they are prepared to exercise restraint. Although such restraint will not ensure success in controlling the spread of nuclear weapons, it is almost certain that without it this danger will never be controlled.

IV The Costs of the MX

15 Dollars and Resources

When President Carter approved the deployment of the MX missile in the racetrack basing mode in September 1979, the total cost in 1978 dollars was estimated to be $33.2 billion. This sum was intended to take care of not only the full-scale development of the MX and the procurement and deployment of 200 missiles with 4,600 shelters but also the operation of the system until 1999, ten years after full deployment. Not included in these estimates were the funds for 2,000 warheads that would be carried on the missiles. The Department of Energy, not the Defense Department, budgets for the nuclear part of weapons systems and that would add a few billion more.

Of this total sum $6.5 billion was for the development of the missiles, $11.0 billion for their procurement, and $10.8 billion for the basing (table 15.1). The approved budget for fiscal year 1981 (FY 81), the first year after the program got underway, authorized $1.568 billion. President Reagan has requested a $2.95 billion authorization for FY 82, about the same amount sought in the final Carter budget. The peak expenditures were expected to occur in FY 84 and FY 85, when the annual rate was estimated to be $8.5 billion (then-current dollars). Although the overall Defense Department appropriations were expected to increase significantly by that time, nevertheless this system could then consume 3 to 4 percent of the total military budget. Even considering the low estimates of 1979, the MX missile system will be a significant item in the national budget and have an even more significant effect on the total funds available for defense. The impact on the national economy is more evident when it is remembered that the Defense budget comprises approximately two-thirds of the total federal "relatively controllable" funds (funds not tied up in statutory programs, such as social security and veterans' benefits and therefore quite easily changed from year to year).

The Carter FY 82, for example, budget had $179.7 billion in controllable outlays, of which $116.3 billion was for national de-

Table 15.1
Estimates of MX system costs

	September 1979 (1978 dollars)	June 1980 (1980 dollars)
Development	$ 6.5 billion	$ 7.65 billion
Acquisition		
Procurement	11.0	12.95
Facility design & construction	10.8	12.94
Simulators	—	1.4
Subtotal of development & acquisition	28.3	34.94
Operations and maintenance	4.9	5.9
Total life-cycle cost	33.2	40.8
Projected cost growth for inflation (76% of baseline to convert to then-year dollars)[a]	—	31.0
Total (then-year dollars)		71.8

[a]U.S. General Accounting Office testimony, June 25, 1979.

fense. The percentage of these funds allocated to Defense undoubtedly will grow in the Reagan administration, and the MX could become an important part of these.

But the ink was hardly dry on these initial estimates of expenditures before the program and its costs began to change. The General Accounting Office in February 1980 analyzed the program and concluded that it could easily reach 60 billion then-year dollars (the dollars taxpayers pay), and it outlined a number of areas where it thought that even these higher estimates were too low. Oklahoma Senator Henry Bellmon, ranking Republican on the Senate Budget Committee, predicted that the cost could reach $120 billion. By May 1980 the air force itself had modified the system and changed its cost estimates. Without including the ten-year costs for operation and maintenance, the development and acquisition costs alone were estimated as $33.5 billion—$5 billion or almost 20 percent more than it had been six months earlier. Despite this increase, which the air force tried to gloss over, the system had been redesigned to effect economies, sacrificing military capability in the process. The separation of the missile and its launcher from the transporter and the decrease in the size and separation of the shelters were estimated to save about $2 billion. These savings, however, resulted in a significant downgrading of the dash capability, which had been called essential in the event that the Soviets saw through our shell game. Expensive dummy missiles had to be added to improve the chances of successful deception. The march toward cost overruns and reduced capability, phenomena almost invariably associated with every large weapons program, was already on its way, only eight months after the decision to go ahead had been made and years away from the groundbreaking in Nevada.

There seems little hope that this trend can be halted in the years ahead. The costs continued to mount during the summer and fall of 1980, all before a single missile had been built or an acre of land chosen on which to base it. In June 1980 testimony by government budget experts before the House Committee on Government Operations, together with new cost estimates released by the air force, provided sufficient data to prepare a new development, procurement, and life-cycle cost estimate of $71.8 billion (table 15.1). By November 1980, little more than a year after the first costs given when the decision to go ahead was announced, the Pentagon's MX office reported that the system would require $108 billion (taxpayer dollars) over its thirty-year life span. This is 500 million dollars per deployed missile. Since only half or 100 missiles are expected to survive a Soviet attack, the costs will be 1 billion dollars per missile available for retaliation. How many more of these skyrocketing cost estimates can we afford? The program is already behind schedule, a factor that always means more dollars. Cost overruns of 50 percent are quite normal for major weapons systems. Where will it end?

These cost estimates were only for the basic system of 200 missiles with 4,600 shelters located in Nevada and Utah. It has become increasingly apparent that this deployment will not be adequate to provide a suitably invulnerable land-based missile system. Defense Department authorities have agreed that in the absence of SALT the Soviet Union could double, triple, or even further increase the number of its warheads. As a response they have proposed to double or triple the number of U.S. MX shelters and add even more missiles. No official figures have been put forth on what funds might be needed for this expansion, but clearly they would be very high. Furthermore, since the program is open-ended, there is no telling when it can be brought to a halt.

If the Soviets continue to add warheads, Defense authorities say we would have to build an ABM system to defend the MX, an entirely new major expenditure. Between 1969 and 1972 we spent about $6 billion to build a totally ineffective ABM system to defend just one Minuteman base. In this case we will need a defense for each cluster, and large numbers of radars and interceptor missiles will be required. At this stage it is impossible to estimate reliably what these ABM costs could be, but Secretary Brown testified in 1980 that this might be $20 billion to $30 billion. Moving some of the MX clusters to Texas and New Mexico or building multiple vertical launch points in Minuteman sites will increase the costs still further. Now the $120 billion estimate made by Senator Bellmon at the time that President Carter approved the MX begins to look optimistic.

Inflation is a key factor in determining what the public will have to pay in taxes for the MX. On the other hand expenditures of these magnitudes can have an important effect on national inflation. Putting $100 billion for weapons into the economy without producing any usable goods is bound to boost inflation. In his message of February 5, 1981, on the economic crisis President Reagan said, "Wars are usually accompanied by inflation. Everyone is working or fighting *but production is of weapons and munitions, not things we can buy or use* [emphasis added]." This is an exact description of the MX. On a nationwide basis this impact might not be more than a few tenths of a percent per year, which although not large compared to the total 12 percent for 1980 cannot be ignored. It should also serve to remind us of the effect of the entire Defense Budget, which is rising rapidly as all other government expenditures are being cut.

Moreover, in certain industries and in some particular regions the impact can be very significant. For example, the air force claims about 2 million tons of cement (probably a very conservative estimate) will be required for the construction on the MX program. The annual total U.S. production of cement is only 85 million tons and in the eleven western states in 1979, cement for existing construction was already being consumed at 99 percent of capacity. Only 200,000 tons were available for new construction. It takes five years to bring a new plant on line. There seems little doubt that the MX will drive the cost of cement up significantly throughout the western states.

Unless very strong steps are taken to establish controls, the inflationary pressures in the MX deployment area will be tremendous. A rapid increase in income in this remote region will produce a strong demand for all types of goods and services beyond those normally available. Prices are almost certain to skyrocket, with real hardship to those who are not profiting from the boom. Except for the relatively unique area of Las Vegas, the regional economy is now delicately balanced. The sudden influx of dollars and the new people moving in can destroy that balance completely. Current wage scales will be completely upset.

The MX program will create a large number of additional

jobs, particularly in California where the missile will be constructed and in Nevada and Utah, the most likely area for the basing. This is, of course, to be expected when there is any major increase in military expenditures. The labor markets in California can absorb this increase without serious dislocation, although there will be a competition for the more technically qualified people between the MX and other military programs. In the deployment area the massive number of new jobs will overwhelm the local labor market, and workers will have to be imported from outside the region.

This increase in jobs in a local area and on a military project is a mixed blessing nationally. A billion dollars of federal funds spent on the MX is a billion dollars of federal funds not available for employing people on other projects. Study after study has shown that Defense programs are poor ways of creating employment. An analysis of the B-1 program made by Chase Econometrics for Rockwell, the B-1 prime contractor, showed that an equal amount spent on public housing programs as on the B-1 would create about 70,000 more jobs over five years. If the money had not been spent by the government at all and instead left to the private sector through a tax cut, 30,000 more jobs would have been created.

David Gold of the Council on Economic Priorities has looked at the MX program in light of these earlier analyses on the employment impact of Defense spending. He estimated that one billion 1972 dollars spent on the MX missile would create a total employment of about 53,000. This same sum would employ 79,000 people if put into mass transit and 65,000 people if spent on solar energy and energy conservation programs. A summary of these analyses is included in table 15.2. Thus there is no question that although the MX would employ large numbers of people, it would not alleviate the national unemployment problem; it would actually make it worse by taking money from other

Table 15.2
Employment impact of MX program (Numbers of jobs per one billion 1972 dollars)

Alternatives	Direct plus indirect employment	Direct employment	Indirect employment
MX missile	53,248	25,055	28,193
Solid waste treatment	65,859	32,173	33,686
Railroad reconstruction	54,220	20,260	33,960
Day care	120,496	103,608	16,888
Solar energy/energy conservation	65,079	—	—
Solar energy	57,235	—	—
Mass transit	79,300	32,500	46,800

programs that would employ more people. In the local areas where employment would be increased, it would be more likely to create labor problems than to solve them.

Furthermore the money spent on the MX would be money not available for other important national programs. There would be less Defense Department money to improve military pay scales, retain trained technicians, procure needed spare parts, buy tanks, and improve readiness. There would be less money to move toward energy independence, a vital security and economic goal. Mass transit, urban renewal, and the reduction of crime—all these and many others would suffer.

Thus the MX program will be a major national venture that will have an economic impact far beyond the regions in which the missiles will be based or built. Although the expenditures for the entire national defense programs will probably be escalating during the next decade, the MX program is bound to be in competition with many other programs for the available funds, technology, and personnel resources. The procurement of conventional weapons, which may have much greater military justification, almost inevitably will suffer if the MX goes forward. A program of this scope is bound to create inflationary pressures not only locally but nationally as well. It will accentuate the increasingly serious unemployment problem by diverting resources from programs with higher manpower utilization. The deployment of the MX is a prescription for accelerating the economic ills that are plaguing us today.

16 Regional Impacts

The scale of the MX deployment program almost defies comprehension. It can be described only by using the superlative case: the biggest, the most expensive, the most in everything except effectiveness. It would be the largest construction project in history. It surpasses the Panama Canal, the Alaska pipeline, even the pyramids. No other project, military or civil, compares with it.

From the start the air force recognized that the effects on the local scene of a project of this scope would be tremendous. Teams were sent into the potential basing areas to survey locations that might be most suitable, and these were evaluated in terms of a set of criteria that needed to be met. A major public relations program was carried out in the Nevada-Utah region in order to convince the public of the requirement for the MX and to reassure them on the consequences. Meetings were held with the local residents to determine the subjects that required study to evaluate the environmental impact. Extensive surveys, including digging wells in search of water, were made throughout the area. Finally, with the assistance of outside contractors, the draft environmental impact statement (EIS), required by law, was prepared and issued on December 18, 1980, about six months behind schedule. It spelled out in great detail the consequences of MX deployment in Nevada and Utah and also in the New Mexico-west Texas area in the event a decision were made to use that location. This draft EIS is the official basis for public review and comments that must be received before preparation of the final statement and before construction can be started in the region. As soon as it was released, it came under strong criticism not only from local inhabitants but from national groups as well. This statement is the basis for most of the background facts used in this chapter.

The Area

If the entire basic system, comprising 4,600 shelters and 200 missiles, were to be deployed in the Great Basin of eastern Nevada and western Utah, the first choice of the air force, the deployment area would encompass a total of about 40,000 square miles (figure 12). This is equivalent to the combined areas of Vermont, New Hampshire, Connecticut, Massachusetts, and Rhode Island. About three-quarters of this would lie in Nevada and one-quarter in Utah. Actual construction would be dispersed over about 15,000 square miles. If it were decided to split the basing and move part of the system to another location, such as west Texas and eastern New Mexico (figure 13), the affected parts of Nevada and Utah would be decreased, but the total area enclosing the two sites would be somewhat larger (perhaps 50,000 square miles, the size of the state of Alabama). Many support facilities, including the command and control centers and aircraft, would have to be duplicated.

The Great Basin is comprised of many broad flat valleys, largely covered with sagebrush. Alongside these valleys are mountains extending more than 10,000 feet above sea level and on whose slopes grow stocky juniper and pinyon trees. In the winter these mountains are blanketed with snow, and the whole region has a great scenic beauty, largely unknown to the inhabitants of the other forty-eight states. It is a lovely recreational area for camping, hiking, hunting, and even fishing. But the environment is very fragile and easily disturbed. A jeep track across the countryside can leave traces visible from the air as long as ten years later. Can anyone even hope that the ravages of the MX will ever disappear from this nearly virgin environment?

Most of the area is public land, managed by the Bureau of Land Management (BLM), but it is nevertheless dotted with small acreages under private ownership. About 9,000 square miles are unsuitable for the basing because they comprise towns,

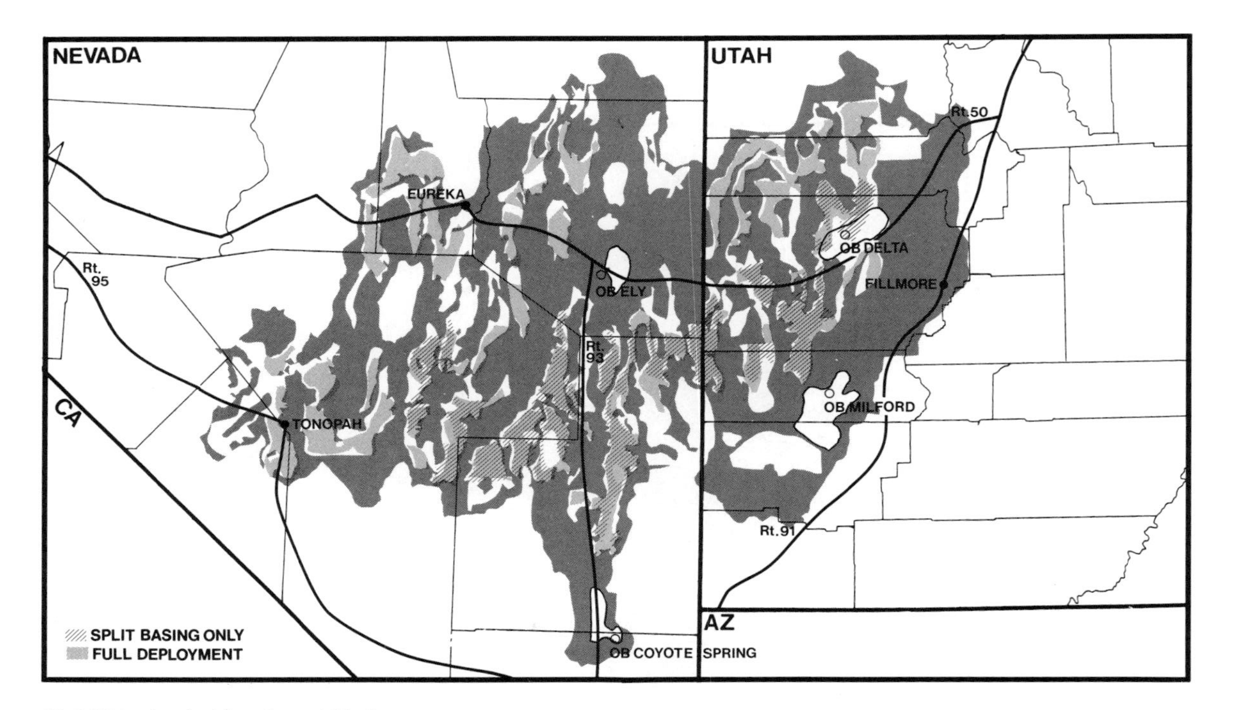

12. MX basing in Nevada and Utah.

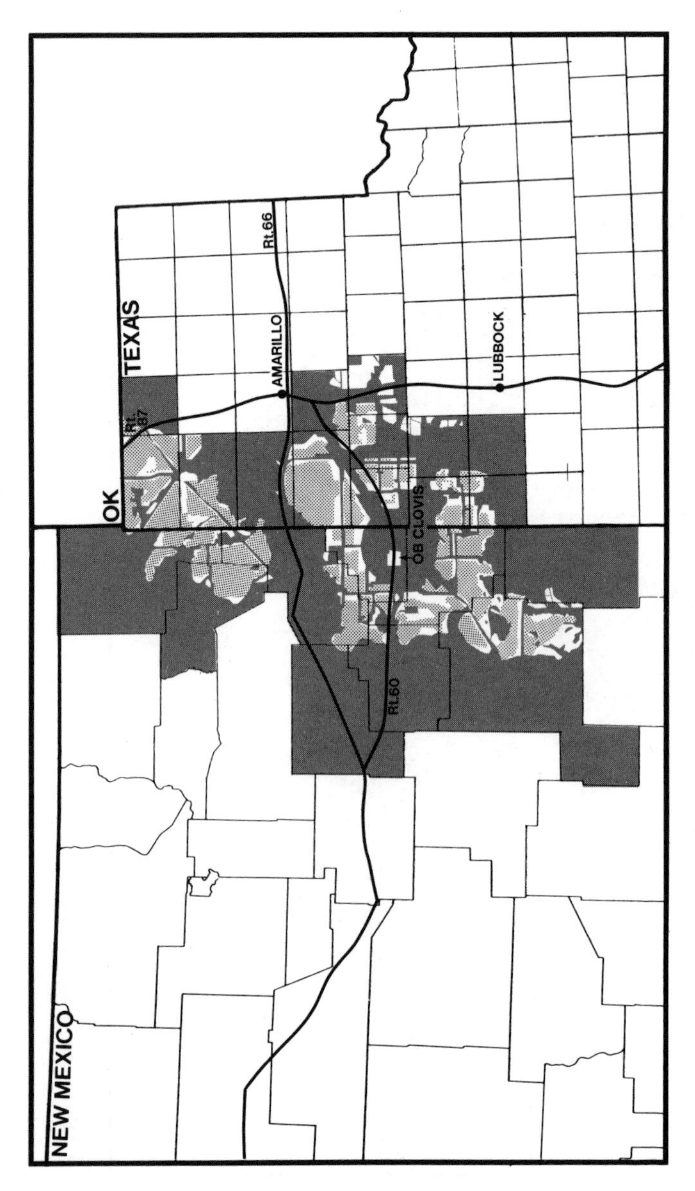

13. Suitable MX deployment area for split basing in west Texas and New Mexico.

parklands, or mountains. The Western Shoshone Indians have a reservation in the middle of the Great Basin, and the tribe has laid claim to 24 million acres, deeded to them under the 1863 Treaty of Ruby Valley. The Indian Claims Commission has been trying to settle this claim for $26 million, which is unsatisfactory to the Indians, and at the moment it is under litigation.

Of the potentially usable 20,000 square miles of flat land, about 12,000 square miles will be needed for the missile shelters in their clusters, maintenance and support bases, remote security surveillance sites, and connecting roads. If horizontal shelters are used in accordance with the Carter plan, the water table must be at least fifty feet below ground, likely in most of the valleys. If the Reagan administration were to move to vertical silos, then the water table must be even lower, and about 40 percent of the valley floors would be unsuitable. Even more land would then be needed. In its December 18, 1980, EIS, the air force stated that if the Nevada-Utah area were to be used, 160,000 acres of desert vegetation would have to be removed.

Area Control

The air force has claimed that only 25 square miles or about 18,000 acres will actually be withdrawn from public use and that the inhabitants of the region need not be overly alarmed. This statement is in the kindest terms deliberately misleading. This minimal area was based on the claim that tight security control would be exercised only over the individual shelters rather than over the entire missile deployment cluster. The air force asserts that security fencing would be placed only within a hundred feet around each shelter and the public would be free to roam over the rest of the land, graze cattle, camp, and prospect for minerals. This completely ignores that the area will be crisscrossed with heavy-duty roads and that large numbers of support facilities, maintenance depots, and constructions camps will be dotted over the landscape, additions not designed to enhance cattle grazing or recreational use. For safety, no habitable buildings will be allowed within about 3,000 feet of any missile shelter.

Even the implication that there would be limited security control is misleading. From time to time, every one of the MX missiles with its ten live nuclear warheads will be trundled from one shelter to another and to the maintenance facilities. The entire effectiveness of the system depends on deceiving Soviet intelligence. Does anyone seriously believe that miners will be allowed to excavate in the area and use seismic and other scientific instruments to locate mineral deposits once the missiles are deployed? Will the ranchers be free to fly their small planes up and down the valleys as they do now in order to get from one part of their land to another? Will hunters in the mountains overlooking the sites be allowed unrestricted use of their guns? No matter what the air force says now, tighter and tighter security measures almost certainly will be imposed. Each cluster is already designed to contain two remote surveillance towers with television cameras and other sensors to monitor activities in the neighborhood, and the valleys will also be dotted with 70 to 120 security alert facilities. Controls will have to be placed on access, and those who pass through will have to be prepared to be searched. It will be a far cry from the present open environment with the freedom to wander in scenic, sparsely inhabited hills.

Roads

Although many facilities must be constructed to service the area, the most prominent features will be the roads. Approximately 8,500 miles of heavy-duty roads will be required to move the missiles from one shelter to another and to support the area as a whole. The cluster roads will be unpaved, treated only with a dust palliative, but the supporting transportation network will be paved. These roads must be capable of handling 1.6 million pound loads—the total weight of the transporter, shield, launcher, and missile. The entire federal highway program, which was begun in 1955, has resulted in building in 25 years about 40,000 miles of roads over the entire United States. The MX program will require building a road system equivalent to about one-quarter as large in this one relatively small, remote region. These will have to be constructed over a period of a few years in locations far from all existing facilities and where water is scarce. The dust that will be kicked up in this desert area will be stifling. Bulldozers, earth movers, graders, and other heavy vehicles will be moving back and forth, raising clouds containing thousands of tons of dust and debris. Can this be controlled by water spraying in a region where every drop of water must be hoarded? The dust palliative proposed by the air force will have to be a miracle chemical.

Support Facilities

National attention has been focused largely on the cluster areas and the construction within them since they are the most dramatic, eye-catching items in the MX system. From the point of view of the impact on the region, however, the support facilities will be of great importance. Each cluster will have within it a maintenance facility for keeping the missiles in operating condition at all times. This cluster maintenance center will serve as a garage for the transporter when it is not in use. Since these maintenance facilities will not be manned under normal circumstances, no personnel need to be housed at these locations. The clusters will, however, have to be enclosed within a security fence, and since they are inside the cluster barrier there will have to be ports in the roof so that Soviet satellites can be sure that extra missiles are not being stored inside.

Because the clusters will be dispersed over a deployment area of approximately 15,000 square miles, it is planned to have three to six area support centers distributed within it. These will be sited so that any shelter will not be more than 65 air miles or 90 ground miles away. These will be relatively large complexes, and approximately 200 to 300 people will be living and working there.

The major permanent support for the MX system will require two operating bases, each covering 4,000 to 8,000 acres and providing personnel and technical support for half of the MX system. These bases will be small cities having a permanent population, including military and dependents, of approximately 17,000 people. If the deployment is entirely within the Nevada-Utah Great Basin, then the air force has proposed that the two bases would be located at Coyote Spring Valley in Nevada and Milford in Utah. The town of Milford now has a population of about 1,200, and the Coyote Spring Valley, about fifty-two miles north of Las Vegas, is currently uninhabited and would require construction of a new civilian support community. It is also surrounded by Bureau of Land Management Wilderness Study Areas, and indeed some of these would fall within its boundaries. The Moapa Indian Reservation is just south of the site, and the base could have a significant impact on the economy, as well as spiritual life, of these people.

The air force EIS indicates that the long-term impact on groundwater at the Coyote Spring Valley site would be high and that water might have to be piped in from outside the region. Cattle grazing in the valley would come to a permanent halt. The Milford base would be situated in an area already experiencing a decline in water level, and the MX use could accelerate this decline. The air force has looked at other potential operating base sites, such as Beryl and Delta in Utah and Ely in Nevada, but the impacts for these sites would be even more severe. The effects from these operating and support bases would be long term since they will continue in operation throughout the life of the MX system. There will be even greater effects during the construction when the peak loads will be very much greater. The MX will also be competing with two coal-fired power plants being built concurrently in Delta, Utah and Ely, Nevada.

Construction

The construction of the MX basing system in Nevada and Utah will be a colossal task, which could tax the resources of any region. In the Great Basin, where there are no resources available, the job will be especially difficult. At a minimum 4,600 large, hardened concrete shelters and thousands of miles of heavy-duty roads must be constructed. The support and operating bases and communities to house perhaps 50,000 people must be built where nothing exists today. More than fifteen construction camps will be required. The air force estimates more than 100 million tons of aggregate must be dug up from these valleys, and 100 to 200 plants will be needed to make concrete. Nearly one-half million tons of asphaltic oil will be required for the roadways. Several marshaling yards, many quarries, and numbers of storage and construction yards will be needed to collect and store these materials. The damage to this sparsely populated and almost untouched area from such scale of construction will be catastrophic. It will not only never be the same again, it will be a unique monument to this generation's mentality.

Dust

Air pollution from the dust that will be stirred up during the construction will be a problem probably unequaled in any other construction project ever attempted. The soil in these areas is very dry and easily kicked up by any kind of vehicular activity. A car traveling across the desert is enveloped in a dust cloud that persists for long periods.

This construction project will be removing the topsoil from 160,000 acres. Bulldozers, earth movers, and other heavy construction vehicles, as well as thousands of small ones, will be crisscrossing the area. No program of dust suppression can be successful under these conditions. The valleys will be in a pall of dust for months on end. Atmospheric inversion conditions can trap the dust clouds in the valleys for long periods. Scenic views as far away as Zion National Park and Bryce Canyon will be spoiled, according to the air force.

The dust may also present a very special and particularly dangerous problem here. In 1959 Kenneth Deffeyes, now a professor of geology at Princeton and then a graduate student, discovered huge deposits of a chemical known as erionite in the volcanic ash layers in Nevada. Erionite, a fibrous mineral similar to asbestos, which is well established as a dangerous carcinogen, is thought to have been the cause of an epidemic of a form of lung cancer, pleural mesothelioma, in some villages in Turkey. The evidence on this danger is not yet all in, but there is a strong possibility that the construction of the MX sites will expose large populations to these erionite fibers and hence to a high risk of lung cancer. Some of the valleys where the clusters will be built have already been shown to contain deposits of erionite fibers in the dangerous-particle-size ranges. The air force paid virtually no attention to this hazard in its EIS. Since this form of cancer typically takes twenty to forty years to develop, we shall not know until it is too late whether the MX has produced a hidden health hazard. In the next century many might regret the air force's oversight in not considering this danger, just as they are now regretting the failure to recognize the dangers of atmospheric nuclear and chemical warfare testing in the 1950s.

Water

The availability of water could be the most critical determinant of whether the MX can be deployed in the Great Basin of Nevada and Utah. Water is the life blood of that part of the country. That parched region averages about eight inches of rainfall annually and has been designated a critical area for water conservation by the Water Resources Council. Shallow wells supply most of the needs at the moment, but the present supply is only just enough, and in some cases not enough, to serve current requirements. Deep underground wells might increase the available water, but no one knows whether this will harm the shallower water supplies so critical to the present way of life.

Wells to tap the deep aquifers could draw down water in the shallower ones since these may be interconnected. When water pumps in nearby Austin, Nevada, sucked the shallow aquifers dry, trees and plants died and the ground subsided. The sand compacted in the water channel; the shallow water will never come back.

The construction of the MX sites will require tremendous quantities of water for the concrete and road construction and to take care of the needs of the thousands of construction workers and their dependents. The air force estimates on the requirements for water have escalated even more rapidly than their estimates of the cost. A few years ago it was 25 billion gallons for construction and twenty years' operation, but in testimony in 1980, the under secretary of the air force, Antonia Chayes, said it was now thought to be 112 billion gallons. By the end of 1980, the air force was saying it could be 190 billion gallons. The construction alone is now estimated to require 42 billion gallons, and this could be a conservative figure.

Because the annual recharge capacity of most valleys of Nevada and Utah is already fully appropriated, the construction of the MX will force the lowering of the groundwater levels, possibly interfering with existing wells and reducing the groundwater flow and quality. The requirements for water in the cluster areas will not be great once construction is completed, so the water supply in most of those areas might recover; however, even a temporary loss of water could permanently affect the ecology of the area. Appreciable areas of irrigated farm and grazing land might disappear forever. Moreover, the long-term water use could exceed the recharge capacity in the neighborhood of the operational bases.

The air force has agreed to follow state laws in applying for groundwater rights and promised the cattlemen not to reduce the existing water supplies their herds now require. But already the ranchers are up in arms. The air force dug many wells less than a mile from existing ranch wells and even used these as a guide for where to dig. They have filed applications for production rather than test wells in these locations, which ranchers believe will hurt their wells nearby. They claim the air force has already violated its written commitments and that the Army Corps of Engineers is ignoring them.

By law the state engineers are responsible for deciding whether to approve the pumping of any groundwater. They must make the final evaluation of the impact on local uses. Recognizing the seriousness of the water situation and aware that the state engineers might refuse their demands, the air force has proposed that the MX system might tap into the Colorado River water supply for at least the Coyote Springs base. But this would then mean that the Colorado River water would not be available for other uses in the region and more remote parts of the country. Southern California and Arizona also claim to have great need for this water source, and we have a treaty to supply some of this scarce water to Mexico. Thus the impact of the MX deployment may well be felt far beyond the limits of Nevada and Utah.

Grazing

The major economic activity in the Great Basin is grazing, mostly of cattle but to a lesser extent of sheep. This region provides one of the largest natural livestock wintering areas in the western United States. More than 650 livestock producers graze their herds here each year, and such ranching is the primary source of income for the inhabitants of the area. Because of the limited water and vegetation, the ranchers need the use of large acreages for economical operation (50 to 150 acres per head of cattle per winter grazing season). Although the air force claims that the clusters will be no impediment to the use of the Great Basin valleys, the thousands of miles of roads, excavation, and other construction activities, which will cause vast forage destruction, will almost certainly make such use a practical impossibility. The range is made up of very fragile but highly nutritious vegetation; large acreages of this will be destroyed, perhaps forever.

One special problem will be created for those whose livelihood depends on the grazing in the Basin—an insignificant weed, halogeton, which grows wherever the topsoil is disturbed. Halogeton is a small poisonous plant that hungry and thirsty animals eat when normal vegetation is scarce. It is a dangerous poison; sheep are easily killed, and cattle get very sick and often die. The weed is a serious problem now with only few roads and minimal activity in this part of Nevada and Utah. Even the MX survey operations have caused serious problems. Imagine how the weed will spread with the vast disruption of the topsoil that full-scale MX construction will entail. General McCarthy, the air force MX representative in the region, when asked what the air force intended to do to control halogeton, is reported to have replied, "We will have the troops pick the weeds if necessary."

The MX construction almost certainly sounds the death knell for grazing on thousands of square miles in Nevada and Utah. In December 1980, the Nevada Cattlemen's Association wrote to President-elect Reagan, "Our livestock industry cannot live with the construction and operation of the land-based MX . . . the livestock industry will be destroyed both in the direct construction area and in a wide band of surrounding areas." The Utah Department of Agriculture has reported that every rancher who grazes cattle in the Utah section of the MX deployment area is opposed to having the missiles in the Great Basin. They fear that the MX construction will force them off the ranges, which they have spent many years and hundreds of thousands of dollars in developing. Range grazing permits, the foundations of the livestock enterprise in the region, are bought in the marketplace. Private ranchers must pay for range improvements, such as fencing, water development, and revegetation.

Sylvia and Dean Baker have a large and well-tended ranch with many acres of irrigated land adjacent to winter and summer range allotments near the Utah border in the midst of the MX deployment area. Dean Baker points out that a balanced range livestock operation is the most energy-efficient method of maintaining a herd of cows and sheep in the United States today. Sylvia Baker has said, "I am convinced that if the MX comes to this area of Nevada, the Nevada livestock producer will face his own demise in the next decade."

Mining

The next most important economic activity in the MX area is mining. Jack Christensen, executive director of the Utah Mining Association, has said the valleys where the air force intends to put the MX clusters contain more than thirty known mineral deposits, including gold, silver, uranium, barium, beryllium, molybdenum, lead, zinc, tungsten, and potash. Many of these we are now forced to import in large quantities. The area contains large coal deposits, which the state envisages using as fuel for huge new power plants. Old mineral deposits are now being reinvestigated because the price of minerals has skyrocketed in recent years and made formerly unprofitable ores worth exploiting.

The MX deployment could slow the mining boom because this industry would have to compete with the MX for labor, materials, and equipment. An air force mineral resources study has revealed that the MX project has already had a dampening effect on industrial exploration for minerals. The necessary controls on operations in the area will certainly also interfere with mining activity. Congressman James Santini of Nevada has expressed dismay that the air force intends to use lands that have good mineral potential.

People

The present population in the Nevada-Utah MX deployment area is about 35,000, approximately one person for every two square miles. The biggest communities in this area in Nevada are Ely and Tonopah, with 6,000 and 3,000 people, respectively. In Utah, Delta with 2,000 people and Cedar City with 11,000, both on the periphery of the MX area, are the largest. Most of the area has small, scattered towns, similar to Baker, Nevada, with only 50 inhabitants.

What will happen if the MX takes over the Great Basin? During the construction period, which could begin in 1982 and reach a peak in 1985, 30,000 people will be directly employed on the MX project. At least that many more will be employed in support, and the population increase for the area could easily reach 100,000 people. The *Las Vegas Review Journal* in December 1980 estimated that the MX could add 250,000 residents to Nevada and Utah. The local citizens, who have enjoyed a stable community life for years, will be vastly outnumbered, becoming a minority in a mass of migrant construction workers from outside the region. The small communities will become boom towns with all the social problems that this will bring. Governor Scott

M. Matheson of Utah has stated, "The influx of a very large number of people into a set of small relatively undeveloped, sparsely populated, culturally homogeneous rural communities will destroy the chosen way of life as it is now known in those communities—forever."

Crime, drunkenness, drug abuse, delinquency, and prostitution will flourish in the boom town economy where there is lots of money and little to spend it on. The strict Mormon ethic, which dominates Utah, will be undermined. These social problems have been demonstrated over and over again in other localities, which have boomed to a far lesser extent than would those in the Great Basin. Dr. Charles Cortese, professor of sociology at the University of Denver, has described an energy boom town, Craig, Colorado, where the population increased 80 percent (not the 400 percent expected in the MX area) between 1973 and 1976. Crimes against property went up 222 percent, crimes against persons rose 900 percent, drug abuse increased more than 600 percent, and family disturbances rose more than 350 percent. "The kinds of things that I am worried about losing," says Laura Cotts of Cedar City, Utah, with reference to the MX, is "being able to send your children across town without worrying."

The air force vows that the social problems can be made manageable with proper community planning. But Stan Albrecht, professor of sociology at Brigham Young University and a specialist on the effects of rapid growth in local areas, disagrees. In a nationally televised debate in Salt Lake City in April 1980 ("Bill Moyers' Journal"), he said, "The benefits are primarily economic that tend to be exported. The costs are primarily of a social nature that tend to remain in the local community, and I very seriously question the ability to adjust adequately to the impact that will be imposed upon them." He said he had never heard of a case where rapid population growth had been handled well by a small community suddenly overrun by the arrival of such large forces. Major Leland Roper of Delta agreed: "It's going to change life here, no matter what we do. Construction workers go from place to place and bring economic and social problems with them. They don't feel welcome. They hustle local women, get drunk, fight."

The boom will present the local governments with formidable and expensive tasks; housing, roads, sewerage and solid-waste disposal, fire and police protection; all must be provided for this population explosion. New schools will have to be built on a vast scale. The air force estimates that 826 additional teachers will have to be hired to staff them, but once the construction is completed, the need for these teachers and educational facilities will drop drastically. Health services will also have to be vastly expanded. The air force admits that it may be difficult to attract the teachers, doctors, dentists, and nurses who would be required only in the peak years. Adequate education and rural health care is already difficult to achieve, but with the MX the problem could become critical.

How will these additional services and facilities be financed? Local governments do not have as wide a variety of state and federal assistance programs as they have had in the past. The future of federal assistance programs to local governments is uncertain in light of the strenuous efforts of the Reagan administration to cut the civilian part of the federal budget. Yet without outside financial aid, significant degradation in services will occur. The air force admits that no local jurisdiction would be able to cope with the MX growth without substantial outside assistance. Federal legislation, which may not be forthcoming today, may be required to mitigate the financial impact.

The construction of the MX system will completely use up the local labor market. Wages will skyrocket as contractors seek to entice workers from remote locations. Local businesses will be unable to compete in this market. As one rancher said, there will no longer be any cowboys available to herd the cattle or mend the fences. Local stores will be unable to supply the needs of the new population. Instead of profiting from the boom, most will be driven out of business by the influx of chain stores from the outside. The costs of all goods and food will rise drastically, producing untold hardships on those who are not part of the boom economy. By the time that the out-of-State workers leave and the bust cycle begins, few local establishments will be left.

Native Americans

The MX deployment in Nevada and Utah will have a profound impact on the Native Americans living in the area. The air force EIS points out that there are over 2.5 million acres of Shoshone, Paiute, Washoe, and Ute Indian reserve lands in the states of Nevada and Utah; sixteen reservations would be within the deployment area. There are some 313 known Native American ancestral and sacred sites, thirty-nine of them within one mile of planned construction activity. The construction of clusters in the Railroad and Little Smokey Valleys surrounding the Duckwater Reservation and the proposed operating base in Coyote Spring Valley have the potential for causing serious water shortages in these Native American communities.

The MX construction will open these ancestral sites to exploitation and desecration unparalleled in the past. As the air force admits, the new access will make it impossible to prevent vandalism of those sites of cultural and religious significance to Native Americans. Mark Bird, a Nevada sociologist, has described the problem to the air force:

> Not only are there likely to be multitudes of non-Indians crisscrossing the reservations, it is probable the nearby construction activities will contribute to the presence of land speculators, missionaries, and bartenders. Historically, each of these occupations has fostered rather lethal consequences for Native Americans. Maintaining pre-construction conditions on reservations may therefore prove as difficult as preserving hemophiliacs in a pool of sharks.

Other Environmental Effects

Many more subtle and less obvious changes will profoundly alter the nature of the area. Moving large numbers of migrant workers into a virtually unpopulated, unspoiled area will affect what Governor Matheson has called "one of the most fragile ecosystems to be found anywhere within the continental United States." The valley fringes and the mountains, prime recreation areas for small numbers of people, will be opened to hoards of hikers, souvenir seekers, campers, hunters, and fishermen—not all of whom will have an appreciation of the land's unique beauty or the knowledge of how to respect and maintain its rare values.

Wildlife ranges will be cut down. The air force has singled out four species on which the impact might be particularly serious: the pronghorn antelope, the sage grouse, a rare desert tortoise, and the big horn sheep. The Coyote Spring operations base will be only ten miles from the Desert National Wildlife Refuge, an area protecting hundreds of these sheep. Several fish species in this desert area will be endangered. The Northern Nevada Native Plant Society questions the continued survival of over a hundred plants if the MX construction proceeds. Native American relics, archeological sites, and paleontologic fossils, which abound in the area, will be lost. Even with maximum care, the ecology and historical resources of the region will suffer tremendously. According to the air force, it will be virtually impossible to control access and use so as to prevent a serious impact on the area.

Expanded MX System

All of the impacts that have been discussed in this chapter and all of the versions of the air force EIS are based on the deployment of a 4,600-shelter system. As Defense authorities point out, the number of shelters, and perhaps missiles too, will have to be increased to 10,000 or 15,000 if the Soviets deploy more warheads than the 6,000 allowed under the SALT II Treaty. The Defense Department is also considering deploying ABMs to defend the MX, which will require still more construction. Under these circumstances the consequences can be further enlarged and last for longer periods.

Impact on Other Potential Basing Locations

Most of the attention on the consequences of MX deployment has been directed to the Nevada-Utah area, since that is the region favored by the air force. However, as a result of congressional pressures, the EIS investigated basing in the west Texas-New Mexico area as well. A major difference between the two locations is that in Texas and New Mexico most of the land is privately rather than government owned and would have to be condemned before it could be used. The costs could be high and delays prolonged. Unlike the Great Basin, many people would have to be relocated. Even the final impact studies could take a long time if the public opposed them. In Utah the contractor doing the research for such studies had to pay a $30,000 fine for trespassing.

Although the air force has considered the case of where the entire MX deployment would occur in the New Mexico-west Texas region, it is more likely that the basing would be split between this area and Nevada-Utah if all of it cannot be put in the Great Basin. The areas that might be involved in such a case are shown in figure 13. This would lighten the burden on each region but would not prevent the impact from being very large in both. Furthermore it is likely to produce a more widespread public and political opposition to the MX deployment. Many people in New Mexico and Texas are already objecting. Governor Clements of Texas, when asked whether he favored locating the MX missile system in Texas, said flatly "No, I do not."

Many of the consequences that have been discussed for

Nevada and Utah also apply to Texas and New Mexico. This area is also very dry, and the need for water for the MX will be in direct competition with the current uses of water in the region. The agricultural losses would be greater than in Nevada. The social consequences of a large population influx will be similar in both areas. More people will have to be displaced to make way for the MX clusters because of the higher population density and private ownership of the land. Only the effects on Native Americans and wild animals will be less severe in the Texas–New Mexico region. Splitting the basing will not make the MX deployment locally acceptable in either region and will only enlarge the dimensions of the disaster.

No environmental study has been started for any type of multiple launch point basing in the Minuteman complexes in Wyoming, Nebraska, Colorado, Montana, or North and South Dakota. Were a decision made to move there, then a long delay could be anticipated before construction could begin. Most of the land in these bases is also privately owned, so additional time would be spent in procuring it. Many of the consequences would be similar to those found for Nevada and Utah. No matter where the air force goes, it will be impossible to construct a missile complex on the scale of the MX project without having a tremendous impact on the neighboring communities.

Public Attitudes

From the statement opposing the MX by the First Presidency of the Mormon Church, May 5, 1981: "Our fathers came to the Western area to establish a base from which to carry the gospel of peace to the peoples of the earth. It is ironic, and a denial of that gospel, that in this same general area there should be constructed a mammoth weapons system potentially capable of destroying much of civilization."

When the plan to base the MX in Nevada and Utah was first

raised in 1979, the local people were concerned about the impact but generally supported the project. In the past the citizens and the political leaders in these two states have been favorably inclined toward military programs. Nevada accepted without any complaints the Nuclear Test Site near Las Vegas, and Utah the Chemical Warfare Proving Ground at Dugway. As accidents have occurred as a result of operations in these military test sites, local skepticism has grown, but still not to the extent of producing uncompromising opposition.

Now in 1981 local hostility to the MX is very strong. Throughout the Great Basin area and its peripheral communities, the citizens are strongly opposed. The true nature of the impact on the region is becoming understood, and the inhabitants see the project as the end to the way of life they sought when they moved there. They have become disillusioned about the national security needs as they learn more about military issues. They have asked questions of Defense Department authorities and have not been satisfied with the answers given them. They are worried at becoming a radioactive sponge for increasing numbers of Soviet warheads.

They were willing to make sacrifices if the safety of the nation demanded it, but now they believe they are only the victims of the bureaucracy. Only in urban Las Vegas, where a small sector of the economy stands to profit financially without paying any of the prices, is the MX welcomed and support for it still rising. Mayor Bill Briare has proposed a high-speed train connecting Las Vegas with the Coyote Spring base to bring workers to the gaming tables. During the daytime, since access is open, it could bring tired gamblers to view the MX clusters where they might place bets on which shelter houses the MX missile.

In the November 1980 election an advisory ballot in eight Nevada counties to be used by the MX showed the people more than two to one against; none of the counties polled had a majority in favor. A Rocky Mountain Poll in February 1981 showed the people in this entire eight-state area were now opposed. These new polls reflected a complete about-face from two years earlier. The MX has united many groups with normally conflicting interests. Ranchers and miners are now working side by side with the environmentalists and the native Americans to stop the program. Governor Matheson of Utah and Governor List of Nevada have expressed opposition to the program. Senators Laxalt and Garn, normally supportive of all military programs, have urged that at least half of the MX deployment be moved outside their states. Except in Clark County (Las Vegas), the political opposition to the MX is large and still growing.

It is not surprising that the people in communities affected by the MX deployment would be hostile to its arrival in their neighborhoods. It is not only that the social, economic, and environmental consequences will be extreme. The MX will also destroy their way of life, will strip away the very reason they moved into these localities in the first place, and force them to start again with their lifetime goals compromised. Senior citizens who have moved to retirement homes in Florida quite properly do not take kindly to being ousted when condominiums are more profitable for their landlords; native New Englanders would not be happy to see the neon lights of Los Angeles County transferred to the valleys of Vermont; and the farmers of Kansas and Nebraska would hate to see the land they love turned into a maze of heavy-duty roads crisscrossing its fertile soil.

If the MX were vital for our future security, then the people in the basing areas would be willing to pay the price, and the rest of the country would be grateful to them for having made the sacrifices. But the MX all too clearly does not satisfy this criterion. Our security will be lower with it than without it. No matter where it is placed or how it is designed, multiple launch point deployment of the MX missile is the prescription only for a social and environmental catastrophe.

V Alternatives to the MX

17 A Sea-Based Alternative

The MX missile system in any of its proposed forms does not solve the problem of land-based ballistic missile vulnerability. The chances that a nuclear war will break out are greater with the MX than without it. And the costs not only in dollars and resources but also in damage to the society and environment, no matter where in the United States and in what form it may finally be deployed, would be enormous. Fortunately there are alternatives; the sacrifices that the MX would require and the risks that it would entail are not necessary.

In searching for alternatives, it is important to remember that crash solutions and decisions are not required. Only one-third of our strategic deterrent is even theoretically threatened in the next decade or more. We still have the other two legs of the strategic triad—the intercontinental bomber force soon to be armed with cruise missiles and, most importantly, the ballistic missile submarine force for which there is no foreseeable threat in this century. Even the fixed land-based ICBMs are vulnerable only theoretically since it is almost inconceivable that Soviet leaders could be confident of destroying them in a surprise attack and of having their society survive. Thus we have time to study carefully what should be done and need not feel driven to continue the MX under any circumstances without considering a change to some other alternative. The concept that we have a window of danger from the early 1980s until the MX is built is only a tactic designed to force hasty, ill-considered, and unwise decisions. It is important that we not commit ourselves prematurely to a program that will be increasingly hard to change, even though it may clearly be unworkable. We need only remind ourselves of the waste had we in 1977 prematurely selected the approach of deploying the MX in tunnels, which would have provided no protection.

At the present time the United States has only begun the full-scale development of the MX missile and has not carried out a single flight test. No land has been transferred to the air force for its basing; we have only made studies, which serve to highlight its serious impact. Although the expenditures to date of a few billion dollars are large in comparison to many nonmilitary programs, it is a small fraction of the total that will eventually be committed if the MX program goes ahead. Now is the time to look carefully at alternatives and avoid irretrievable actions that could tie our hands in the future and lead to dangerous reactions abroad.

In view of the dismal outlook for designing a survivable land-based ICBM system in the absence of long-term constraints on the Soviet force levels, we should seek alternative ways to broaden the base of the strategic deterrent. These alternatives should meet the objectives that led to the decision to deploy the MX in the first place. The criteria that should be considered in the search for alternatives are summarized as follows.

1. Any deterrent weapons must not only be survivable but nontargetable. If they can be located and aimed at, they cannot forever remain survivable. Unless these criteria are satisfied, the inevitable result will be a continued arms race with uncertain prospects for achieving security. Survivability must not depend on successful deception of the Soviet Union, for this is the wrong game to be playing and the Russians are the wrong opponents. Furthermore, reliance on deception will inevitably result in a situation where we can never be confident that we have been successful.

2. The size and characteristics of any alternative system must be verifiable by national technical means, perhaps with the assistance of some agreed nonintrusive arrangements, for we must not set a precedent for secret deployments that in the long run can only work to the Soviet advantage.

3. The weapons should have good, tight command control characteristics so that they cannot be launched without a positive order from the top. Two-way communication between those

who have custody of the weapons and command authorities is important.

4. Matching the Soviet Union should not be a criterion in selecting any new weapons system. To allow the Soviets to determine the weapons we procure is a mindless approach to our security. Our technologies are different from theirs, and we usually are more advanced; our geopolitical situation is different, and our strategies should be different. We should have different weapons instead of being driven by a desire to match the Soviet Union in every category.

5. Our future weapons systems should not have a first-strike capability against any major element of the Soviet deterrent because this can only invite a dangerous reaction on their part. New weapons should be able to attack a limited number of hard targets at most, but they should not be able to threaten the entire Soviet ICBM force.

Moving more of our strategic deterrent to sea is the best way of satisfying these criteria. One proposal was to use a buoyant missile and distribute it around the ocean or even inland waterways with surface ships. This would have the advantage of providing an infinite number of launch points rather than a finite number of targetable shelters. If they were to be dispersed long in advance of their use, however, they might be detected by satellites. On the other hand if they are kept in surface vessels until just before launch, then the ships become potential targets. It might be possible to target all ships that are potential distributors of these floating missiles. Furthermore Soviet intelligence would always have the potential to find out which ships carried the missiles; then we would be back to playing the shell game, this time with ships in place of shelters.

This type of deployment would make it impossible to verify the number of missiles available for launching. There would be no way to count the number stowed in surface vessels. Finally, this system would have the disadvantage of leaving large numbers of live missiles with nuclear warheads floating around the oceans out of direct U.S. control. While they presumably would be recoverable, there would always be a danger that one would be lost or picked up by some hostile group.

Submarines are much better carriers for missiles. They are survivable, and at the present time there is no foreseeable Soviet threat that could destroy a large number of our submarines in a surprise attack. Since their location at any particular moment is unknown, they are an ideal mobile, nontargetable platform for carrying and launching missiles. Unlike aircraft they can remain at their operational stations for long periods. The number of submarines can be verified by satellite observation of their construction and fitting-out facilities. Such technical intelligence methods have worked quite satisfactorily for about ten years under the SALT I Interim Agreement.

In 1967 the STRAT-X study investigated the concept of using small submarines operating relatively close to our coasts as platforms for launching missiles. This idea was not given serious consideration by the Pentagon; instead the United States went ahead with its large Trident submarines, strategic cruise missiles, and the MX land-based missile program. As analyses increasingly demonstrated that none of the proposed ways of making land-based ICBMs survivable was practical, the concept was revived, largely due to the efforts of Sidney Drell of Stanford University and Richard Garwin of IBM. They are both long-time advisers to the White House and the Defense Department on strategic and other weapons matters and thoroughly familiar with the technologies involved. This concept is still not endorsed by the hierarchy in the Pentagon but at the moment shows far more promise of meeting any need for future strategic systems than do others that have been proposed to date.

Such a minisubmarine missile system would comprise a fleet of approximately fifty submarines, each carrying two ballistic missiles for a total of one hundred survivable missiles, the same number that the land-based MX system is designed for. These would be deployed in the near coastal waters off the United States at all times. If there were a firm requirement to have fifty submarines continuously ready to launch their missiles, additional submarines, perhaps as many as twenty-five, might have to be procured to replace those in port for maintenance. If the submarines were designed to carry four missiles each, then the number of submarines could be cut in half and still have one hundred survivable missiles. Fewer submarines, each with more missiles, might lower the cost, but it would reduce the number of target submarines with which Soviet antisubmarine warfare forces would have to cope. There will be thus a trade-off between the system cost and vulnerability.

The missiles would be enclosed in a neutrally buoyant capsule mounted horizontally external to the pressure hull. In order to fire a missile (figure 14), the capsule would be released, tilting up as it rises to the surface. When the capsule breaks through the ocean surface, the missile booster is fired in much the same way as with our existing submarine-launched ballistic missiles. The MX missile could be used if desired, but it would appear unnecessary to develop an entirely new missile for this purpose. The Trident I missile, already available, could be equally well adapted for this mission at a very significant saving in funds and with an increase in flexibility for our strategic forces. This proposal has not received sufficient attention due to strong political pressures for the acquisition of large missiles with countersilo capabilities. If the Reagan administration is really serious about cutting budgets, this should be thoroughly investigated.

If each submarine were to carry two missiles, it need have a pressure hull no larger than 1,150 tons; if it were decided that four missiles should be carried on each submarine, then a 1,500 ton pressure hull would be required. (Our Trident boats dis-

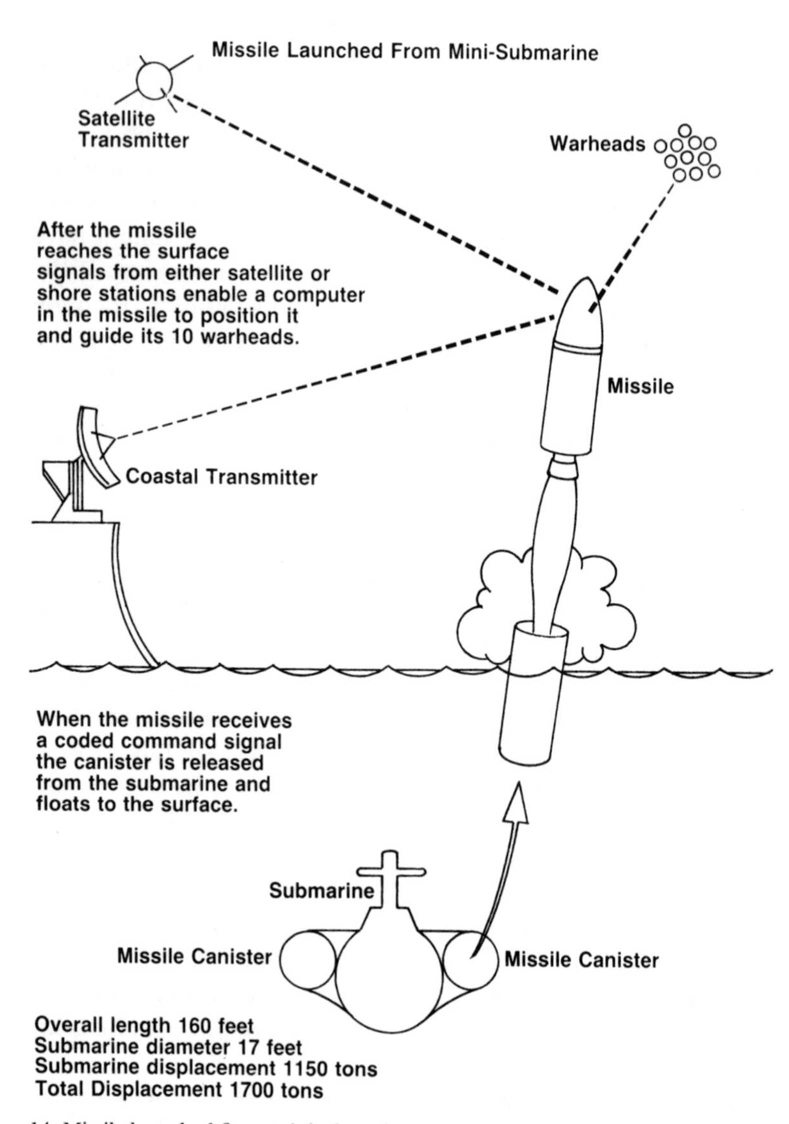

14. Missile launched from minisubmarine.

place 18,000 tons.) The size of the crew should be small, perhaps fifteen to twenty-five persons. Since the submarines will be operating relatively close to the coasts and have no requirement for traveling at high speed, they need not, and should not, be nuclear powered. Instead fuel cells, which have been extensively tested, are capable of driving such a submarine with very low noise levels without any need for surfacing or snorkeling. Some safety problems associated with handling the fuel cell reactants at sea have been postulated, so some research and development work needs to be undertaken on this problem. If these problems cannot be resolved to meet the desired time schedules, the first-generation propulsion could employ commonly utilized diesel electric power.

The cost for such nonnuclear power plants would be very small. If strong supervision is exercised over this program, to ensure that no unnecessary frills are added, the cost for the entire system can be kept low. High-technology equipment will not be needed on the submarine itself; sophisticated inertial navigational systems would not be required on the submarines, which could rely on the missile guidance systems to locate their positions.

The primary function of the crew would be to monitor equipment, to perform routine maintenance and safety checks, and to maintain control over the missile systems. The crew could be rotated about every three to four weeks to reduce the stress on individuals; two or more months of underwater duty is currently required in the Polaris-Trident submarines. If the maintenance required for the missiles is not too extensive, the submarine eventually might not have to return to base for each crew change, which instead could be performed from tenders. Such an operation could be carried out without giving away the location of the submarine when on station.

The minisubmarines would operate in a coastal ocean band approximately 500 miles wide. In order to avoid their becoming vulnerable to a tidal wave that might be produced by a large Soviet underwater nuclear explosion, they should be deployed in waters not less than 400 feet deep. In the Pacific they could be relatively close to shore because the ocean bottom falls off very rapidly near that coast. In the Atlantic, however, the continental shelf extends out some distance. Nevertheless the 600-foot depth line is still less than 100 miles from shore so the submarines would not have to move very far out to avoid becoming vulnerable to a tidal wave produced by nuclear explosion. This 500-mile ocean band would provide an operating area of about 2 million square miles, ample space in which to hide and avoid becoming vulnerable to possible barrage attacks from thousands of Soviet warheads.

The first and most important question that must be asked about this minisubmarine missile system is whether it will be truly survivable into the next century. At the present time there is no foreseeable antisubmarine warfare threat to submarines operating close to the American coasts. Even our larger boats can maneuver not only in the far reaches of the Atlantic and Pacific oceans but even relatively close to Soviet shores with little fear of being detected.

Quietness is the primary determinant of detectability. The smaller minisubmarines should be less vulnerable than existing submarines because their electric drive propulsion can be even quieter than the nuclear propulsion of the most modern Tridents. When fuel cells are available to provide power, the submarines should be still quieter than with diesel power, and this is already quieter than nuclear power. The minisubmarines will also operate at much slower speeds than the Polaris or Trident, which must move long distances to get on station. This again will reduce the detectability of the minisubmarines since the higher the speed of a submarine moving through water, the more underwater sound it generates.

The smaller surface area of these new submarines will also

make it somewhat more difficult for them to be detected by antisubmarine warfare vessels using active acoustic systems (systems in which a sound wave is transmitted through the water and the submarine detected by picking up the acoustic signal reflected from the submarine's hull). Furthermore all acoustic detection techniques, the only methods effective at the present time, are much more difficult to use in relatively shallow water (less than 10,000 feet) than in deep water because of the complex nature of the reflections of the sound waves from the ocean bottom and surface. Thus in every way the minisubmarines should be much more difficult to detect than our larger, existing ones, and even these are virtually undetectable now.

It is unlikely that this situation will change significantly for the foreseeable future. Although predictions more than a decade in advance should always be made with caution, no scientific breakthroughs can be anticipated that might make such minisubmarines vulnerable in the next ten to twenty years. Satellites cannot be used for locating submarines as long as they are beneath the ocean waves. In September 1980 the chief of naval operations, Admiral Thomas Hayward, objected strongly to statements made in support of the land-based MX by some Defense Department officials that our submarines might be vulnerable in the next decade. He felt that these unduly downgraded the survivability of our existing and planned submarine deterrent force, which navy officials said would remain invulnerable through the 1990s. Secretary Brown later supported the admiral's view that the Soviet antisubmarine threat had been exaggerated, because the Soviet technology in this area is far inferior to that of the United States.

The survivability of the minisubmarine missile fleet is a quite different problem from that of a submarine fleet that must operate in wide ocean areas far from home base. The minisubmarines can operate from many bases and will be stationed in waters relatively close to the U.S. coasts, where they can be much more easily protected and where it will be much more difficult for any Soviet antisubmarine warfare force to maneuver. The Atlantic Ocean in a sense is a NATO preserve, which Soviet submarines or antisubmarine warfare ships, must enter through the restricted, Western-controlled Greenland-Iceland-United Kingdom passages. For Soviet antisubmarine warfare ships to operate effectively close to our shores would be very difficult, especially since we can bring our counterforces to bear.

At the present time quietness is the only counter-antisubmarine warfare tactic that our submerged submarines have found necessary to adopt. Acoustic or sound waves are the only way to detect submarines under water. If at some time in the future a plausible threat should appear on the horizon, then many other countermeasures could be considered. Decoys could be developed with the same acoustic characteristics as the minisubmarine. Or the ocean waters in which a submarine is operating could be filled with underwater sound so that the high general noise level would further conceal any noise from the submarines themselves. This can be done relatively easily. A small high-explosive charge is enough to make the waters of the entire Atlantic Ocean reverberate for a long time.

Finally the minisubmarine deterrent force would have approximately fifty of these ships at sea at all times. This is nearly double the number of our existing missile submarines at sea, a very significant increase in the number of targets that Soviet antisubmarine forces must deal with. For the Soviet Union to develop and deploy a capability for destroying in a near-simultaneous attack fifty small submarines deployed within a few hundred miles of our coasts is an almost impossible task. Not only do they have to be detected and tracked continuously, but they must then be destroyed. Doing this from surface vessels or submarines would involve a tremendous Soviet shipbuilding program, which would take years. The United States would have ample time to take countermeasures, making this job even harder.

Sinking the submarines with long-range ballistic missiles is equally difficult. The submarines would still have to be detected and their location continuously known in order to direct the incoming warheads to the proper location. These missile warheads would have to be capable of penetrating beneath the ocean surface in order to be able to destroy the submarines at a sufficiently large distance from the explosion. Surface detonations have a much smaller kill radius than underwater ones. While it is always risky to predict technological advances twenty to thirty years into the future, missile submarines are a good bet to survive; there are so many factors in their favor. There is little doubt that a missile on a minisubmarine has a far greater chance of surviving than does a missile in any of the proposed land-based deployment modes. It is a better gamble to rely on a missile at sea than a missile on land.

Because the location of the minisubmarines will not be known at any specific moment, their missiles, unlike the land-based MX, will be untargetable. Their vulnerability will not be dependent on the number of Soviet warheads, and their deployment will not provide any inducement for the Soviets to increase the size of their force. Improved accuracy in Soviet missiles will not equate with a greater threat; it will only be a wasted asset. The size of the minisubmarine missile force thus is not open-ended as is the MX. The number of missiles needed for retaliatory purposes can be determined in advance, and hundreds of extra missiles need not be procured to hedge against those that might be destroyed in a Soviet first strike.

For these reasons, the day the first minisubmarine goes to sea with its missiles, the United States will have enlarged its survivable deterrent. Every additional submarine will continue to improve it. By contrast in the case of the multiple launch point basing schemes, no significant improvement in survivability is obtained until the number of operational launch points has become comparable to the number of Soviet warheads.

Furthermore the number of missiles deployed on such minisubmarines can be verified using current national technical means. Although the number of submarines deployed at sea cannot be accurately known at any specific time, the number produced and made ready to go to sea can be counted. This is the procedure that has been used for verifying the SALT I ceilings on submarines since 1972, and before that for our intelligence needs. This approach has worked well. We have known for many years the number of Soviet missile submarines and the number of launch tubes each contains. Although the minisubmarines will not require as extensive construction facilities as are needed for the larger nuclear-powered ones, our technical intelligence should be able to detect any significant secret production of such ships.

The joint U.S.–U.S.S.R. Standing Consultative Commission established by SALT I worked out procedures for making it still easier for satellites to see any submarine before it is ready for action. There is no reason why these methods cannot be applied to the minisubmarines as well. The size of the submarine alone would put an upper limit on the number of missiles it could carry. Thus unlike the situation of the MX being deployed on land, the size of the force cannot be augmented rapidly by stockpiling missiles and moving them into already available launchers. No increase in the uncertainty over the size of the threat would develop if the Soviet Union should decide to follow in our footsteps. The minisubmarine missile system would thus be useful if either a treaty were agreed to or an unconstrained race were to continue.

Even if the MX with warheads having a countersilo capability is used as the missile carried by the minisubmarines, the system as a whole would not of necessity become a first-strike threat against the Soviet ICBM deterrent. Furthermore since they would not be targetable, the missiles cannot provoke a Soviet first strike. If fifty submarines were deployed, each carrying two MX missiles with ten warheads, the total force would comprise only 1,000 warheads, not a sufficiently large number, no matter how accurate, to threaten the entire Soviet ICBM force of 1,400 missiles. Even if twenty-five extra submarines are needed to compensate for those normally in maintenance status and these were added to the ready force, this would still be too small to threaten by itself all Soviet ICBMs. The Soviets could be concerned about these 1,000 to 1,500 warheads if they were added to the 1,000 to 1,500 countersilo warheads on our Minuteman IIIs. On the other hand we should be phasing out the potentially vulnerable Minuteman missiles on a one-for-one basis concurrently with the buildup of the invulnerable minisubmarine missiles. The security gain from keeping the total number of our countersilo warheads below the level that can threaten the entire Soviet ICBM force should be kept constantly in mind as the submarine fleet builds up and the Minuteman missiles are withdrawn.

Unlike the land-based version, only 100 MX missiles are needed in the minisubmarines since these are nontargetable and no extras need to be deployed at one time to allow for the 50 percent that might be destroyed, a contingency that must be faced with any land-based deployment. A 100-missile minisubmarine force serves the same military purpose as 200 missiles hidden in multiple shelters; in both cases 100 missiles will be available for retaliation. We should not procure 200 missiles for submarines just because that was the plan for the land-based system. If we act with wise restraint, we can avoid putting the Soviets in the position of having to respond by adopting dangerous strategic postures, such as being prepared to launch on warning or to launch a preemptive strike in times of crisis.

The current Poseidon submarine fleet has a very effective communication, command, and control system, which uses redundant channels to ensure that it is reliable. Because the minisubmarines will be stationed relatively close to the mainland, it is possible to have even better two-way communication between them and the central command authorities. This submarine system will provide opportunities for direct cable links, using large numbers of submerged buoys for the submarine to tie into, or for very low frequency communication with a myriad of land stations and aircraft. This should reduce concerns that have been voiced over current systems.

The minisubmarine's command communications should be as reliable in a combat situation as that for the multiple protective shelters in Utah and Nevada. In the latter case we cannot rely on land lines to provide the orders for targeting and launching the missiles for the lines can be cut. Continuously airborne communication and control centers are planned. The desired flexible response will be difficult when half the land-based missiles may have been destroyed. These problems are less severe with the minisubmarines.

Similarly the proximity of the launch area to land makes it feasible to give the missiles the same accuracy as the MX launched from land. The precise location of the launch point can be provided to the guidance computer by sending data to the missile once it is above the water. Guidance corrections can be supplied by the NAVSTAR-Satellite Global Positioning System or, if the NAVSTAR satellites are destroyed, by line-of-sight radio signals from a large network of inexpensive onshore beacons. Thus if it is desired to be able to destroy a few hard targets other than ICBM silos, then the minisubmarine system can do this. In every respect the minisubmarine can have a weapons capability equivalent to that of the land-based version.

The important differences are that the minisubmarine will be survivable to matter how many accurate warheads the Soviets have and that it will be launched from sea, not from U.S. territory. Some Defense officials have argued that a positive feature of a land-based missile leg of the deterrent triad is that if it is to be neutralized, it will require Soviet attacks on U.S. territory, thus making some American response more credible. However, the people in the United States and particularly the West might disagree and prefer a target area in the oceans rather than having their lands becoming a sponge to absorb Soviet warheads.

It is too early to have definitive estimates of the costs of a minisubmarine system. In the spring of 1980 the navy analyzed a 200-missile minisubmarine system and concluded that it was unlikely to be cheaper than the MX. This analysis, however, was based on fifty larger submarines carrying four missiles each and having larger crews and many additional characteristics not necessary to perform the basic deterrent task. Most important it projected base systems appropriate for training and support of large Trident nuclear-powered submarines. Drell and Garwin have estimated cost savings of $10 billion over the land-based systems with the same number of survivable warheads.

Because the minisubmarines are not targetable, the system does not require enlargement if the Soviets add warheads. Its survivability is independent of the number of Soviet warheads since they cannot attack the submarines. The air force is already planning expansion of the basic MX system, and many of the construction costs are escalating. Thus even if the rather pessimistic navy estimates were correct, the minisubmarine system would almost certainly be cheaper in the long run. It is important, however, that no frills or extra missions be developed for the submarines. Finally, there will be no environmental or social consequences resulting from the deployment of the minisubmarines offshore; the communities in Nevada and Utah, Texas and New Mexico, or Wyoming and North and South Dakota will be spared the disastrous impact of multiple launch point basing for the MX.

The minisubmarine system has been criticized because it supposedly erodes the policy of maintaining a triad of strategic delivery vehicles. The diversity in strategic forces provided by the triad of Poseidon-Trident submarine-launched ballistic missiles, land-based Minuteman ICBMs, and the long-range bombers soon to be equipped with cruise missiles has proved an important element in the security of our deterrent as a whole. Now that the land-based missile component is becoming theoretically vulnerable in the 1980s, that leg is shaky, but certainly there is no cause for panic because the other two components of the triad provide a more than adequate deterrent. What is important is to have a survivable deterrent as a whole. Our past decision to procure this triad was a sound one.

But there is nothing sacred about our triad of strategic weapons, one leg based at sea, one leg in the air, and the third on land. The important factor is to have a diversification of strategic forces so that if one type of weapons system becomes vulnerable, the remaining types can still provide an adequate deterrent. The minisubmarine system does supply that diversification; it has many of the desirable characteristics of land-based missiles and also presents different antisubmarine warfare problems from the Trident and Poseidon submarines if the Soviet Union should seek to neutralize them. The minisubmarine forces provide larger numbers of targets, are quieter and smaller, and will be located in different operational areas closer to our coasts where their protection is easier and a Soviet attack much more difficult.

The only threat to the sea-based legs of this new deterrent triad would be Soviet ballistic missile defenses, which are now banned by the ABM Treaty. If we were to abrogate this treaty in an attempt to shore up the obsolescent land-based missile component, then we would only reduce our confidence in both the sea- and land-based elements of the deterrent. But land-based missiles must not be allowed to become an excuse for destroying this existing treaty, which now enhances our security.

Instead new arms-control measures could be designed to increase confidence in the long-term survivability of these submarines. We might seek an agreement with the Soviet Union for making a belt 500 miles wide along our coasts into a zone in which all antisubmarine warfare activities were banned. A similar zone would have to be established for Soviet boats, perhaps in the Barents Sea and the Sea of Okhotsk. Such an agreement could be adequately verified, since significant threats to these submarines in these waters could not be put in place covertly.

This minisubmarine missile system would give us a much more effective, diversified deterrent than were we to insist on maintaining the potentially vulnerable land-based leg of the deterrent. Former Secretary of Defense Melvin Laird and former chairmen of the Joint Chiefs of Staff, Admiral Thomas Moorer and General Maxwell Taylor, have all recommended that we not deploy the MX on land and that we move our deterrent to sea. The unstated reason why this most promising alternative has not been given greater acceptance is its lack of support within the active Defense Department bureaucracy. To drop land-based missiles and move to sea would cut the air force out of any long-range strategic missile program. The navy is reluctant to adopt the minisubmarine program because eventually it might be viewed as an alternative to the very large Trident nuclear-powered submarines. The minisubmarines have no doting parents to guide them in their infancy through the turbid bureaucratic seas. In the meantime our security, our tax dollars, and our people, particularly in the West, are being made the victims.

William Perry, President Carter's under secretary of defense for research and engineering, has claimed that the abandonment of land-based ICBMs in favor of going to sea would be widely perceived around the world as acquiescing to Soviet dominance. Such a view is a good example of how the misuse of perceptions can be employed in weapon decision making when valid technical arguments are lacking. A U.S. move of its strategic missiles to sea should properly be perceived as conferring on the United States an advantage because three-fourths of the Soviet deterrent would be left potentially vulnerable while virtually our entire deterrent would be made invulnerable. To allow misperceptions of real military strength and interservice rivalry to dictate our security decisions is only a prescription for insecurity.

18 Limiting the Threat

The conference table, though scarred by many past frustrations, cannot be abandoned for the certain agony of the battlefield. Disarmament, with mutual honor and confidence, is a continuing imperative. Together we must learn how to compose differences, not with arms but with intellect and decent purpose.

From the Farewell Address of President Dwight D. Eisenhower, January 17, 1961.

The only satisfactory long-term solution to the maintenance of a secure stable deterrent is to reduce and eventually eliminate the danger of nuclear war. This can be done in two ways. The first is to reduce the threat through arms limitations, achieved through either negotiated agreements or reciprocal national actions. The second is to get rid of land-based ballistic missiles so that there will be no target for the aggressor.

Trying to find some gimmick to make land-based ICBMs invulnerable is a losing game in the long run. New ways of basing new weapons on land will never be lastingly satisfactory. Technology, particularly in guidance, has advanced to the point that the aggressor will always have the advantage. Any weapon that is targetable, that can be located and attacked, can never be truly survivable and can only lead to a race between offense and defense with at best uncertain success for the party on the receiving end. Basing weapons at sea is more satisfactory since it avoids many of these difficulties, but this will always be costly. Although there is no foreseeable danger to submarine-based systems, there will always be worries in the decades ahead that antisubmarine warfare eventually will become a threat. Opportunities will grow for competition between submarine technology and antisubmarine warfare. The only road to stable security in the next century is through arms limitations.

For success through arms control the Soviet forces must be reduced to the point that they would not have sufficiently numerous accurate warheads to threaten our entire ICBM force. But if such a lowered force level is to be negotiable, be stable, and endure, then it is essential that the reverse situation also exist. The lack of threat to ICBMs must be mutual; the United States must not have sufficient numbers of accurate warheads to threaten the Soviet ICBM force. Although this concept is somewhat hard to accept, it is probably more important for U.S. security that Soviet ICBMs do not appear threatened than that U.S. ICBMs are not threatened. This anomalous situation occurs because only one-third of our deterrent is at stake when ICBMs are vulnerable, while three-fourths of that of the Soviet Union would be at risk. We are not safe when the Soviet leaders think the bulk of their nuclear deterrent is in jeopardy.

SALT II was a step in the direction of limiting the threat to land-based ICBMs, but it was too late and too little to solve the problem. It placed an upper limit on the number of MIRVed ICBMs (at 820 missiles) and also a restriction on the number of warheads each missile could carry, but these ceilings were too high. It would have allowed the Soviets about 6,000 ICBM warheads, more than enough to threaten our 1,052 ICBMs in their single silos. We were allowed 550 Minuteman III missiles with three warheads and more than 200 MX missiles with ten warheads, a threat to the 1,400 Soviet ICBMs. No controls in SALT II were placed on accuracy improvements because of the difficulty of verifying that such restrictions were being abided by. Even had the ceilings agreed to by President Ford at Vladivostok in 1974 been put into immediate effect, they would also have been too high to ensure survivability of either nation's land-based ICBM force; in fact they were higher than the eventual SALT II ceilings. The Carter proposal of March 1977 set somewhat lower limits, but even the 550 MIRVed ICBMs in that offer would have allowed each side to have a countersilo threat. The opportunity to control this threat was lost earlier by the failure to address the MIRV problem seriously in the SALT I negotiations.

Nevertheless the SALT II Treaty provisions would have been useful first steps; they would have started the process and improved the climate for the possible future achievement of low enough ceilings. Eventually the number of warheads would have to be reduced to fewer than 3,000, less than half the SALT II level. The notion that the procurement of the MX missile will succeed in pressuring the Soviets into stopping and cutting back its weapons programs is misguided. That was the course that got us into the current predicament. That bargaining chip tactic was a failure with MIRVs in 1970. Now it will only ensure still-larger Soviet deployments in the future.

Limitations cannot be solely on Soviet weapons programs if we are to reach this goal of eliminating the threat to ICBMs. The essential characteristic of any arms limitation agreement is that it must serve the mutual interests of both sides. The United States cannot expect the Soviet Union to make large reductions in its land-based ICBMs, 75 percent of its strategic force, while the United States proceeds unrestrained. This was one of the major drawbacks of President Carter's March 1977 proposal, which otherwise had some very useful provisions for limiting dangerous new weapons but obviously looked one-sided to Soviet eyes. In the future we must be willing to put controls on the MX program if we wish the Soviets to forgo their countersilo missiles. We cannot expect to proceed with the deployment of unlimited numbers of ground- and sea-launched cruise missiles while Soviet ballistic missiles are cut back. A halt to the testing of all new missile systems without loopholes for any "sacred" programs and a low quota on all types of missile tests are examples of future arms control measures designed to slow dangerous improvements in missile systems and to reduce reliability, which is needed if a first strike were ever to be contemplated.

To be realistic, however, the prospects for arms control agreements with the Soviet Union to eliminate the mutual threat to ICBMs are currently not good. The climate for agreement on any subject is very cold. The concept of linkage of arms control agreements with Soviet good behavior around the world is not likely to hasten SALT III, which could provide nuclear arms limitations that would be in our security interest in the midst of the cold war. Several years may pass before meaningful negotiations can even begin and many more before concrete results can be achieved. In the meantime both nations will face each other with increasingly vulnerable ICBMs, a situation that can only enlarge the risks that a nuclear conflict will break out. National actions, hopefully but not necessarily precisely reciprocated, can be very valuable. We should take to heart the views of Lord Mountbatten, former chief of the British Defense Staff and chairman of the NATO Military Committee, who said on May 11, 1979, three months before he was assassinated: "The real need is for both sides to replace the attempts to maintain a balance through ever increasing and ever more costly nuclear armaments by a balance based on mutual restraint. Better still, by reduction of nuclear armaments, I believe it should be possible to achieve greater security at a lower level of military confrontation."

Once we understand that the existence of countersilo missiles is detrimental to our own safety, then we should recognize the wisdom of halting the modernization of our ICBMs with warheads having a countersilo capability. Just as we are quite willing to take on our own actions to acquire new weapons that improve our security, so we should be prepared to take similar actions to withhold deployments that decrease our safety. The procurement of more ICBMs that can only become beacons to attract a Soviet attack reduces our safety. We should be prepared to refrain from such deployment. Although we might hope that the Soviet Union would reciprocate, we would still be better off than we would otherwise even if they do not. Categorizing such a halt to this dangerous procurement of countersilo weapons with the pejorative term *unilateral disarmament* demonstrates ostrich mentality when it is in our security interest. We have to stop placing our safety in jeopardy by succumbing to such politically inflammatory slogans. Misleading perceptions of weakness must not become a path to nuclear war.

Although the prospects for progress in SALT are bleak at the present time, other arms control measures could contribute to strategic stability and simultaneously serve our mutual goal of slowing the spread of nuclear weapons to other countries. The first of these would be a comprehensive test ban treaty, halting the development of new nuclear warheads. Although both nations have warheads available for use in all their new strategic systems, a test ban would make more difficult the acquisition of a silo-busting force. For example, if the Soviets desired to put twenty or thirty warheads on their SS-18, more testing might be needed to get a small nuclear explosive with a high enough yield to threaten our land-based ICBMs. Similarly a treaty halting the production of fissionable material for weapons could prevent the Soviets from having enough plutonium for 10,000 to 15,000 MIRV warheads. Measures such as these are verifiable and would improve our security. We should seek to negotiate them now because they are in our interest; we should not wait until we are satisfied with Soviet political behavior, by which time they could have all the warheads that they need. Finally we should place controls on strategic antisubmarine warfare in order to increase long-term confidence in the survivability of submarine-based strategic missiles.

As a second approach toward reducing the instability from vulnerable ICBMs, we should work toward the gradual removal of ICBMs as targets. Strategic weapons that are targetable are remnants of earlier eras. In the 1960s and 1970s ICBMs in fixed hardened silos could not be destroyed because technology was not up to the task. But by the 1980s technology has advanced to the point that they have outlived their usefulness, in fact have become a liability. Pouring more good money after bad to try to retain the attributes of the past is foolish. Instead we must look for other ways to maintain a secure deterrent and to maintain the desirable, but not the undesirable, characteristics of ICBMs. There is nothing sacred about a land-based ballistic missile as such for a leg of the triad. That was just an accident of the technology of that time. We should begin thinking about phasing out land-based ICBMs. Then there will be no targets for the Soviets to aim at or attract their fire in a time of crisis. Any money they spend on acquiring a countersilo capability will have been wasted and not spent on items of real military value.

Survivability of the deterrent and the ability to assure unacceptable retaliation under any circumstances were the goals in the 1960s, and they still should be the goals today. If we can maintain these goals by means of arms control, either through negotiations or through national restraint, then we will have maintained our security and reduced the risk of a nuclear catastrophe. This is the prescription for security now and in the next century.

Glossary

accuracy
The ability of a warhead to hit near its intended aim point, usually measured in terms of the probability of hitting within a distance (CEP) from that point.

acoustic submarine detection
The use of sound (acoustic) waves to detect and track submarines in the seas. Active systems rely on sending out an acoustic signal and receiving its reflection from an object such as a submarine; passive systems rely on receiving the sound emitted by the submarine itself.

active defense
The protection of civil and military targets through the use of defensive weapons, such as surface-to-air missiles, antiballistic missile missiles, and antiaircraft artillery. See passive defense.

antiballistic missile (ABM) system
A system for destroying ballistic missiles; normally consists of radars and interceptor missiles to detect, track, and destroy offensive missiles.

antisubmarine warfare (ASW)
The detection, identification, tracking, and destruction of hostile submarines. ASW can be either strategic (aimed at neutralizing an opponent's ballistic-missile submarines) or tactical (concerned with the pursuit and destruction of submarines in a local situation for missions such as convoy defense and aircraft carrier defense).

aquifer
An underground water-bearing rock formation that can be used as a water source.

ballistic missile
A missile, classified by range, that moves on a free-falling trajectory under the influence of gravity. Much of its trajectory will be above the atmosphere.

bus
The final stage of a missile that is used to dispense MIRVs on trajectories toward their targets.

circular error probable (CEP)
A measure of missile accuracy. It is the radius of a circle around a target in which 50 percent of the missile warheads aimed at that target will land.

civil defense
See passive defense.

cluster
A group of shelters connected by roads and from which a single MX missile could be launched.

command, control, and communications (C^3)
The system for authorizing and transmitting to the missile the command to launch. With nuclear weapons this now originates with the President and passes through military command channels until the missile firing mechanism is activated.

countersilo capability
The ability of a missile warhead to destroy a hardened ballistic missile silo.

cruise missile
A pilotless plane that flies through the atmosphere, very often close to the earth's surface. It can be launched from aircraft, ship, and submarines, and from the ground.

dash capability
The ability of a missile to move rapidly from one launch point to another when warning of an attack has been received.

disarming strike
See first strike.

dummy
An object to replace the MX on its transporter so that Soviet intelligence will not know at which shelter the missile has been deposited. It must be able to simulate all the characteristics of the MX missile itself, such as weight, emitted radiations, and heat. If ABMs are also decep-

tively deployed to protect the MX, then they will also have to simulate dummy MX missiles.

fallout
The spread on to the earth's surface of radioactive particles from clouds of debris produced by nuclear blasts. "Local" fallout falls to the earth's surface within twenty-four hours of the blast.

first strike
An initial attack with nuclear weapons. A disarming first strike is one in which the attacker attempts to destroy all or a large portion of its adversary's strategic nuclear forces before they can be launched. A preemptive first strike is one in which a nation launches its attack first on the presumption that the adversary is about to attack.

fission weapon
A weapon deriving its energy from the splitting of heavy atoms, such as uranium or plutonium; sometimes called A-bomb.

fixed land-based missile
A missile that can only be launched from a single predetermined location.

fratricide
The destruction or degradation of the accuracy and effectiveness of an attacking nuclear weapon by the nearby explosion of another attacking nuclear weapon. This phenomenon would decrease the effectiveness of an attack on closely spaced targets, such as missile silos.

fusion weapon
See thermonuclear weapon.

hard target
A target protected against the blast, heat, and radiation effects of nuclear explosions. Hardening is usually accomplished by means of earth and reinforced concrete and is measured by the number of pounds per square inch of blast overpressure which a target can withstand.

implosion
High compression of bomb materials to permit fission or fusion reactions to take place.

intercontinental ballistic missile (ICBM)
A land-based ballistic missile with a range greater than 4000 nautical miles.

kill probability (k_p)
The probability that one missile will destroy another.

kiloton (kt)
A measure of the yield of a nuclear weapon, equivalent to 1,000 tons of TNT.

launch-on-warning
A strategic doctrine under which a nation's bombers and land-based missiles would be launched on receipt of warning (from satellites and other early-warning systems) that an opponent had launched its missiles. Sometimes recommended for use when there is uncertainty over the ability of fixed-site ICBMs to survive an attack.

megaton (mt)
A measure of the yield of a nuclear weapon, equivalent to 1,000,000 tons of TNT.

MIRV (multiple independently targetable reentry vehicle)
A missile with MIRVs can carry several warheads, each of which can be aimed at separate targets.

mobile missile, horizontal
A missile that can be moved in a horizontal attitude from one location to another for launching.

mobile missile, vertical
A missile that can be moved in a vertical attitude from one location to another for launching.

multiple launch point basing
Many launch points, i.e., shelters or silos for firing a land-based ballistic missile in order to force the enemy to attack many targets in order to destroy a single missile.

multiple protective shelters
A form of multiple launch point ICBM basing.

mutual assured destruction
A concept of reciprocal deterrence which rests on the ability of the two nuclear superpowers to inflict unacceptable damage on one another after surviving a nuclear first strike.

national technical means
The use of technical intelligence collection means for verifying compliance with negotiated arms control agreements. These means must be consistent with the recognized provisions of international law.

NAVSTAR
A global positioning system for providing course data to ballistic missiles and other weapon systems. NAVSTAR consists of twenty-four satellites in synchronous orbits providing near-continuous signals to passive receivers aboard ballistic missiles, allowing them to determine their positions within twenty to thirty feet and to correct their attitude or velocity.

on-site inspection
A method of verifying compliance with an arms control agreement whereby representatives of an international or other designated organization, or of the parties to the agreement, are given direct access inside national boundaries to view force deployments, weapon systems, or tests.

passive defense
The protection of civil and military targets without the use of weapon systems. Measures for protecting population (evacuation, blast shelters, fallout shelters) and economic targets (dispersal, hardening, redundancy) are ordinarily termed "civil defense." See active defense.

payload
The weight of the weapons or any penetration aids carried by a missile. See throw-weight.

polar trajectory
A ballistic missile trajectory crossing the polar regions; for example, a trajectory between the U.S. and the U.S.S.R.

pounds per square inch (psi)
A measure of nuclear blast overpressure or dynamic pressure, used to calculate the effects of a nuclear detonation or the ability of a structure to withstand a nuclear blast.

preemptive strike
See first strike.

recharge capacity (water)
The ability of a water supply to build back to its original level after it has been drawn down for some use, such as building MX bases.

reconnaissance satellite
A satellite containing cameras, radio receivers, or infrared detectors to gather information on the deployment and characteristics of the weapons of other nations. A most important element of national technical means of verifying arms limitations and determining the nuclear threat.

reentry vehicle
That part of a ballistic missile designed to re-enter the earth's atmosphere in the terminal portion of its trajectory.

reliability
The probability that all elements of a missile system will perform according to their designed parameters.

SALT
Strategic arms limitation talks, negotiations between the U.S. and the U.S.S.R. to limit strategic weapons. SALT I began in 1969 and ended in 1972 with the ABM Treaty and the Interim Agreement on certain offensive weapons. SALT II began in 1972 and a treaty was signed on

June 18, 1979. This has not yet been ratified. President Reagan has talked of going directly to SALT III.

second strike
A follow-up or retaliatory attack after an opponent's first strike. Second-strike capability describes the capacity to attack after suffering a first strike. The strategy of deterrence is premised on high confidence in the ability to deliver a nuclear second strike that would inflict unacceptable damage on the nation that struck first.

selective ABM defense
An ABM defense that only fires its interceptor at an incoming warhead when it has determined that the warhead will destroy the target, that is, an MX missile.

silo
A cylindrical hole in the ground from which a ballistic missile can be launched; normally it would be hardened to make it resistant to blast pressure and other effects of a nuclear explosion.

submarine launched ballistic missile (SLBM)
A ballistic missile that can be launched from submarines; the range can vary from 600 to 6000 nautical miles.

survivable weapons
Weapons that can not be destroyed by an enemy strike.

targetable (non) missiles
Missiles whose location can be determined at a specific time and thus are subject to attack. If they are nontargetable, then they do not draw enemy fire.

thermonuclear weapon
A weapon deriving its energy from the fusion reaction of light elements, such as the hydrogen isotopes, tritium and deuterium; sometimes called H-bomb.

throw-weight
The useful weight that a ballistic missile can place on a trajectory toward a target. It includes the warhead, any penetration aids to counter

an ABM, and the mechanisms for dispensing and guiding the reentry vehicles in the case of MIRVs.

triad
Strategic forces that are composed of three parts: land-based intercontinental ballistic missiles, submarine-launched ballistic missiles, and long-range bombers.

verifiable weapons
Weapons whose number and type can be verified by national technical means or other agreed procedures.

yield
The force of a nuclear explosion expressed as the equivalent of the energy produced by tons of TNT. See kilton and megaton.

Many of these definitions were obtained from "A Glossary of Arms Control Terms" prepared by The Arms Control Association.

Index

Numbers in italics refer to figures.